Design Analysis of Support Insulators for HVDC Gas Insulated Switchgear

Authors

Dr. P. Kowstubha, Dr. M. Balasubbareddy, Dr. A. Supraja Reddy and Dr. P. Sathish

Integrated Publications TM
New Delhi

Published By: *Integrated Publications* ^{TM}

Integrated Publications
H. No. - 3 Pocket - H34, Sector - 3,
Rohini, Delhi-110085, India
Email – info@integratedpublications.in

Authors: *Dr. P. Kowstubha, Dr. M. Balasubbareddy, Dr. A. Supraja Reddy and Dr. P. Sathish*

Edition: *1st*

Publication Year: *2023*

Pages: *73*

ISBN: *978-93-5834-003-7*

Book DOI: *https://doi.org/10.22271/int.book.324*

Price: *₹445/-*

Abstract

Gas Insulated Switchgear (GIS) for High Voltage Direct Current (HVDC) requires solid insulators to support the conductors mechanically and to separate pressurised gas compartments. The existence of the spacer has an impact on the flow of SF6 gas insulation, particularly where the solid insulating spacer and SF6 gas meet. To reduce the internal and surface discharges of the insulator, the electric field levels on it must be controlled. Most of the failures in GIS system are due to improper design of insulator both in terms of profile and dimensions. Thus, the shape of the spacer plays an important role in withstanding electrostatic field levels. The spacer is to be designed to realize uniform electric field distribution on its surface.

This book presents a thorough analysis of the variables influencing the distribution of the electric field on the surface of the spacer when a cone insulator is considered. Further, the effect of static charges on the electrostatic field distribution is also analysed. Numerical software is used to carry out electrostatic analysis of the HVDC GIS spacer. The optimization analysis is carried out on cone epoxy insulator at different operating conditions to arrive at optimal dimensions of spacer with appropriate profile. The study primarily aimed to analyse various factors that affect GIS for HVDC applications.

Contents

S. No.	Chapters	Page No.
1.	**Introduction**	**01-04**
	1.1 Introduction	01
	1.2 Geometrical Model of the Cone Insulator	03
	1.3 Objectives	04
	1.4 Outline of the Book	04
2.	**HVDC Gas Insulated Switchgear**	**05-13**
	2.1 Introduction	05
	2.1.1 Present Scenario	06
	2.1.2 Future Developments and Future Trends in HVDC GIS Technology	07
	2.2 Advantages of HVDC	07
	2.3 Insulation in HVDC GIS	07
	2.4 Insulator Design and Estimation of Electrical Stresses for HVDC	08
	2.4.1 Solid Insulation	08
	2.4.2 The Insulator Shape	08
	2.4.3 Cone and Post Insulators	09
	2.5 Electric Field	10
	2.5.1 Electric Field Stresses	11
	2.5.2 Uniform and Non-uniform Electric Field	11
	2.5.3 Applied Voltage	11
	2.6 Surface Charges	12
	2.7 Insulation Properties of GIS Insulators under Lightning Impulse with Superimposed DC Voltage	13
3.	**Electro 2D Software**	**14-27**
	3.1 Introduction	14

	3.2 Assigning Physical Properties	14
	3.3 Defining the Type of Model	15
	3.4 Balanced and Unbalanced Rotational Symmetric Models	15
	3.5 Setting the Material Defaults	15
	3.6 Adding Materials	16
	3.7 Assigning Boundary Conditions	16
	3.8 Assigning Surface	17
	3.9 Generating 2D Elements	18
	3.9.1 2D Elements and the BEM Solver	18
	3.9.2 Boundary Element Method (BEM) in Electric Field	18
	3.9.3 Principle of Boundary Element Method	18
	3.10 Creating Graphs	19
	3.11 Static Analysis	22
	3.12 Transient Analysis	22
	3.12.1 Setting the Time Variables in Transient Analysis	23
	3.12.2 Types of Transient Source	24
	3.12.3 Standard Signal	26
	3.13 Running Transient Simulation	26
	3.14 Analysis of Transient Simulation Results	26
4.	**Electrostatic Field Analysis Using Electro 2D**	**28-36**
	4.1 Introduction	28
	4.2 Simulation	28
	4.3 Analysis of Cone and Post Insulator	28
	4.4 Steady State Analysis	29
	4.5 Transient Analysis	29
	4.6 Methodology	30
	4.6.1 C2D Model Creation in Electro	30
	4.6.2 Defining Material and Boundary Condition	31
	4.6.3 Defining Element and Simulation Run	33

5.	**Simulation Results**	**37-59**
	5.1 Results of Static Analysis	37
	5.1.1 Case 1: V and E Distribution with NO Insulator using Static Analysis	37
	5.1.2 Case 2: V and E Distribution with Post Insulator using Static Analysis	39
	5.1.3 Case 3: V and E Distribution with Cone Insulator using AC Analysis	41
	5.2 Results of Transient Analysis	43
	5.2.1 Case 1: E-field Distribution with NO Insulator Using Transient Analysis	43
	5.2.2 Case 2: E- field Distribution with Post Insulator using Transient Analysis	47
	5.2.3 Case 3: E- field Distribution with Cone Insulator using Transient Analysis	51
	5.3 Comparison of Electric Field Stress with NO Insulator, Post Insulator and Cone Insulator	56
	5.3.1 Steady State analysis	56
	5.3.2 Transient Analysis	57
	5.4 Comparison of Static and Transient Analysis	58
6.	**Surface Charge Accumulation in HVDC**	**60-68**
	6.1 Introduction	60
	6.2 Factors that Affect Surface Charge Accumulation	60
	6.3 Analysis of Surface Charge Accumulation	60
	6.4 Methodology	61
	6.4.1 2D Model Creation in Electro	61
	6.4.2 Defining Material and Boundary Condition	61
	6.4.3 Defining Element and Simulation Run	63
	6.5 Results of Surface Charge Accumulation	64
7.	**Conclusions**	**69-70**
	References	**71-73**

Chapter - 1

Introduction

1.1 Introduction

Due to the rapid increase in energy consumption under the constraints of urbanisation, gas insulated systems have been widely used since the late 1960s. SF6 gas, an insulator, a conductor, and an outer enclosure make up most of the the system's parts. To provide mechanical support for the conductor in a gas-insulated busduct, solid insulators are used in GIS. They serve the duty of providing the space between the high voltage electrode and exterior enclosure in addition to providing mechanical support. Since significant flashovers indicate the collapse of these supporting structures known as insulators, accurate modelling of the insulator is crucial for power engineers.

The goal of the ideal spacer design is to increase the high voltage Gas Insulated Switchgear's (GIS) mechanical and electrical performance. One of the important parts of a GIS is an epoxy spacer, which serves as a supporting conductor, a way to divide it into separate components, and an insulator between the conductor and the enclosure. Electrical safety as well as mechanical reliability of epoxy spacer should be taken into consideration [2] due to the requirement of compact-sized GIS and greater gas pressure in GIS.

Gas Insulated Switchgear (GIS) has been utilised extensively in power transmission lines because to its excellent reliability and safety. Epoxy spacers, which serve as supporting conductors and electrical insulators, play the most important role in the electrical insulation performance of GIS because the demands and needs of electrical equipment in power transmission lines are compact and operated at higher rated voltage [3] [4]. Many studies on SF6 alternative GIS have been conducted recently, and it is well recognised that most alternative gases have inferior insulating capabilities than SF6 gas. The gas pressure of an alternate gas should be higher than that of SF6 gas to address this weak point without expanding the overall size of GIS [5]. To prevent cracking and delamination, that eventually contributed to the collapse of the GIS's insulation, epoxy spacers were designed with dependable mechanical strength. The design of insulation for power equipment has seen extensive use of numerical field calculations. To enhance the insulation performance, an electric field optimisation based on the computation was created [6].

The development of high voltage direct current (HVDC) technology, which offers more benefits for the transmission over long distances, for the integration of renewables, and for connecting asynchronous networks, is currently trending, despite the fact that high voltage alternative current (HVAC) technology is still the most popular for the transmission of electric energy [7].

Two-dimensional cone epoxy insulator is designed for 325KV HVDC Gas Insulated Switchgear is used to calculate electric field stress and surface charge as constraint of an optimal design.

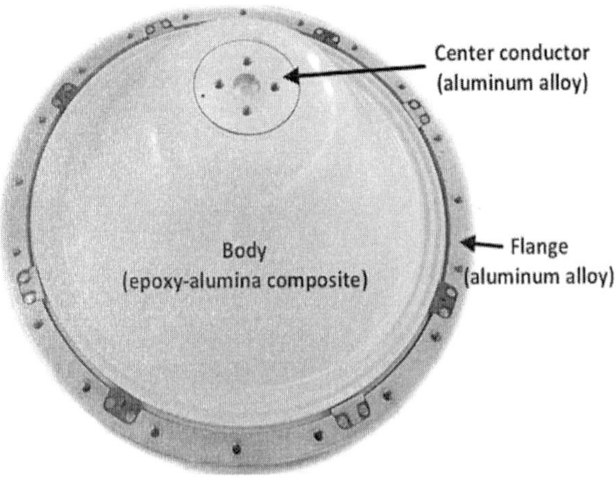

Fig 1.1: Structure of a UHV GIS spacer

To realize compact and highly reliable HVDC equipment, the following parameters with respect to insulators are mainly considered.

1) Insulation properties of the insulator's surface and its gas gap
2) Design of insulators for direct current (DC) applications
3) Characteristics of charge accumulation on the insulator surface and techniques for measuring them
4) Measures for particle trapping and particle movements,
5) HVDC applications for partial discharge detection and analysis [7].

Advantages of GIS

GIS offer economic solutions in mountainous terrain and sites of hydroelectric power stations where the site preparation costs are high.

1) As GIS are hermetically sealed units, snow at high altitudes does not affect the performance of the substation equipment.

2) GIS are useful in urban and industrial settings where space and pollution are the key concerns and useful in coastal areas with salt pollution.

3) With reduced area requirements and reduced height of GIS equipment, seismic considerations and associated civil engineering works do not pose serious problems for substation design.

4) The GIS use modular designs, which give the substation layout a great deal of flexibility and make the best use of available space. As a result of these substations' typical closer proximity to the load centre, transmission and distribution network losses are decreased.

1.2 Geometrical model of the cone insulator

Fig. 1.2 shows the geometric model for electrode configuration using Cone shape epoxy spacer, SF6 gas, Enclosure and HV conductor.

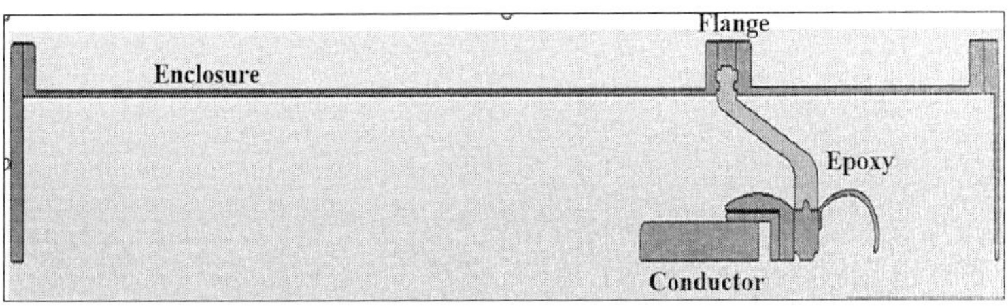

Fig 1.2: Cone epoxy spacer using HVDC analysis

The GIS has a stainless steel or aluminum sealed metal casing that maintains the gas under pressure and prevents SF6 leakage. Due of its small dimensions, it can be positioned outside or within a building. Although steel is also frequently used, aluminum is typically cast or welded for enclosures. An enclosure made of metal tubing has high voltage conductors positioned in the centre. An insulator holds up the conductor. Conductors are made of pure aluminum and high potential is given to the conductor for the transmission of electrical energy. Shield designed for better field distribution and to avoid localized stress on any path. Shield has finger contact, which is made up of copper chromium and it is used for the transfer of current. Flange is made up of aluminum which is given zero potential voltage. GIS enclosure modules are connected to one another using bolted flanges that seal the enclosure with "O" rings. The GIS modules' individual metal enclosure pieces are electrically connected to one another. The effectiveness of the electrical insulation of the GIS is most significantly impacted by the epoxy spacer, which serves as a supporting conductor and electrical insulator. These support insulators are composed of epoxy resin, which is carefully cast to prevent the development of voids or cracks during the curing process.

1.3 Objectives

1) Formulation of problem geometry.
2) Designing of cone epoxy insulator and to optimise the electric field stress.
3) Static analysis for no insulator, post insulator and cone insulator in HVDC GIS.
4) Transient analysis for no insulator, post and cone insulator and find electric stress in HVDC GIS
5) Detailed comparison of static and transient analysis of different geometric configurations in different conditions.
6) Effect of space charge accumulation on the insulator in HVDC GIS.

1.4 Outline of the book

This book is documented into seven chapters, brief description of which is given as follows:

Chapter 1: Gives an overview of the book, literature survey and outlines the main objectives of the thesis.

Chapter 2: Presents a detailed description of HVDC Gas insulated switchgear, insulators, surface charge.

Chapter 3: Detailed description of 2D Electro (version 9.2) Software is presented.

Chapter 4: Electrostatic field Analysis using Electro 2D Software is presented.

Chapter 5: Demonstrates the results of electric field stress using Static Analysis and Transient Analysis.

Chapter 6: Analysis of surface charge accumulation is presented.

Chapter 7: Presents conclusion of the work.

Chapter - 2

HVDC Gas Insulated Switchgear

2.1 Introduction

The present high voltage network must be extended all the way up to the consumer due to the rising demand for power and the increasing energy density in major cities. At the substations close to the actual users, stepping down the voltage from transmission to distribution level not only produces economic benefits, but also guarantees a steady supply of electricity. Due to these factors, a brand-new kind of switchgear has been created that takes up the least amount of space possible while yet meeting all current criteria and being environmentally friendly.

Several GIS units are still in use after more than 30 years of operation. They don't have many lengthy bus routes. GIS and circuit breakers utilise over 80% of the world's SF gas production. Excellent dielectric strength, current interruption capabilities, and high heat transfer capacity are all characteristics of SF6 gas. For a secure and reliable supply of electricity, gas-insulated switchgears are placed as substation equipment and reception equipment. The use of a high insulating performance gas is used to seal metal containers containing circuit breakers, disconnectors, earth devices, etc. Switchgear plays a crucial role in any network of power systems. Switchgear aids in the smooth operation of electrical appliances from power generation to distribution and safeguards against overload and short circuit damage.

High voltage direct current (HVDC) technology offers more benefits for the transmission over long distances, for the integration of renewables, and for connecting asynchronous networks, even though high voltage alternative current (HVAC) technology is currently the most used for the transmission of electric energy.

To boost transmission power and improve trunk network system stability, a variety of high-voltage gas insulated switchgears (GIS) and gas insulated circuit breakers (GCB) for AC and DC voltages have been created. The reliability of DC insulation and its design were frequent problems in this evolution. Other problems with HVDC include the fact that higher voltages for HVDC transmission necessitate thorough understanding of insulating system behaviour owing to DC-application.

Important material characteristics, like resistivity, vary greatly and fluctuate with temperature. All these impacts must be taken into consideration when designing internal insulation. Additionally, there are common types of electric stressors that are entirely distinct from AC-fields, such as static state. Resistivities, which can vary from one material to another by several orders of magnitude, are what control DC fields.

When designing insulating insulator for HVDC GIS, insulation characteristics such as charge accumulation on insulator and resistive electric field at HVDC voltage application have to be considered which differ from the characteristics for HVAC GIS. This analysis deals with charge accumulation phenomena on insulator, to study the Electric field characteristics around insulator under HVDC fields and the optimum configurations of insulator for HVDC GIS. Two-dimensional cone epoxy insulator is designed for 325KV HVDC Gas Insulated Switchgear is used to calculate electric field stress and surface charge as constraint of an optimal design.

In addition to UHV AC transmission, there is growing need for high-voltage DC (HVDC) installations that are dependable and affordable, particularly for long-distance power transmission. It is commonly known that insulating materials behave differently as a dielectric under DC stress than they do under AC or impulsive voltage stress.

For an optimised design and for the dependability of HVDC equipment, knowledge of the DC behaviour of insulating gases as well as solid materials like insulators is crucial. In order to improve stability in trunk network systems and enable increased transmission power, various high-voltage gas insulated switchgears (GIS) and gas insulated circuit breakers (GCB) for AC and DC voltages have been created recently. The reliability of DC insulation was a problem that plagued this development on a regular basis. DC insulation is crucial for the equipment design in HVDC GIS and GCB.

2.1.1 Present scenario

Since the introduction of SF6 to industry, GIS companies have installed thousands of indoor and outdoor GIS worldwide. Research and continuous development of the first system types paved the way to the current generation of GIS units. They are among the most compact and most reliably performing designs available, used internationally by such renowned companies as ABB, Siemens, Hitachi, and Mitsubishi. With a low weight, long operating life, low operating costs, and modular, space-saving design, GIS can be extremely economical.

2.1.2 Future developments and future trends in HVDC GIS technology

GIS technology is constantly changing and evolving. There is an increasing effort to improve GIS design, manufacturing, testing, and operation technologies and best practices. High reliability of GIS is the main factor, which contributes to its worldwide use. Future trends in the development of GIS should be focused on alternative clean gas to SF6, compact switchgear design using three-phase modules, use of vacuum circuit breaker cells in the medium-high voltage GIS and fewer breaks per pole in high voltage circuit breakers, and optimisation of GIS design to allow for easier maintenance. The appropriate management of spare parts and the creation of methods for the efficient use of outdated equipment are additional significant issues facing the sector.

2.2 Advantages of HVDC

Since the first half of the 20th century, DC transmission technology has been developed and improved. However, it is now, particularly in light of the energy transition, becoming more and more significant. It is frequently the best option to satisfy today's demands for integrating renewable energy sources, connecting power systems with various parameters, or delivering big amounts of energy over great distances. The advantages of HVDC being are

1) Interconnection of energy systems.
2) Lower power losses during long distance transmission of electricity.
3) Building energy networks through maritime areas.
4) Power output from renewable electricity generation areas.

2.3 Insulation in HVDC GIS

The GIS insulation system is a mixed insulation, consisting of both SF6 and solid insulation materials. Besides the insulation properties of the gas, the behaviour of the solid insulation has to be considered. Normally support insulators are made of epoxy resin filled with a filler like alumina or silica material. There are different designs for insulators regarding shape and inserts for electrical field steering.

1) For the design of insulators, the following points must be considered
2) Electrical strengths, field distribution

3) Mechanical strengths and gas-tightness, especially for barrier insulators separating the gas compartments.

4) Sensitivity against particles.

5) Resistance against decomposition products. Insulators with alumina fillers have very good behaviour in this respect.

6) Free of partial discharges. Each insulator is subjected to a very sensitive partial discharge measurement in the factory during the high-voltage power frequency test.

2.4 Insulator design and estimation of electrical stresses for HVDC

2.4.1 Solid insulation

In a GIS, conductors are insulated from the enclosure by the gaseous medium as well as the supporting solid insulating insulator. The maximum design electrical stress in the gas gets reduced because of the lower gradients of the solid insulation.

2.4.2 The insulator shape

The amount of field stress placed on the insulator is significantly influenced by its form. It is anticipated that a suitably designed insulator will offer an electrical field stress that is less than the ionisation initiation. The electric field stress on an insulator is significantly influenced by the profile of the weather shed and the end fitting designs. The maximum electric field strength was found to be in the vicinity of the triple junction, according to research by Chakravorti and Steinbigler [15] into the relationship between a porcelain-shaped post-type insulator and the limit of electric field strength that surrounds it in the presence and absence of pollutants. This study looked at elements such the insulator's axial height, core radius, electrodes' radius, and the weather shed's slope angle [15].

In designing Insulators, following points are considered:

1) Insulator axial height and peak electric field intensity are inversely proportional to each other.

2) The maximum field intensity is barely impacted by the size of the core radius.

3) The field intensity is only marginally affected by an increase in slope angle.

4) The electric field intensity was significantly decreased when the shed radius was extended by as little as 6 cm.

5) When there are no pollutants present, an increase in the radius of

the electrodes will result in a rise in the field intensity for an insulator and a decrease in the field intensity once the insulator surface is contaminated.

Additionally, it has been shown that the shield's design has a considerable impact on the field around the live end. In [15], Phillips designed and simulated three distinct types of insulator fits in the absence of a corona ring. According to their findings, a shield's rounded edges are always associated with a reduction in the field intensity that is greatest close to the shield. Additionally, it was found that the field distribution in the vicinity of the larger-diameter end fitting was more aesthetically pleasing.

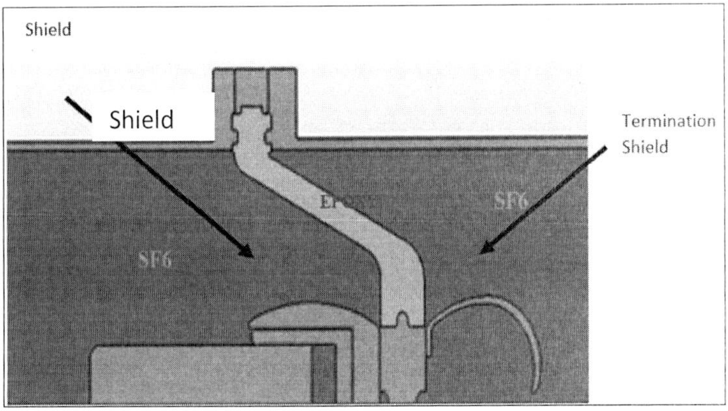

Fig 2.1: Shape of insulator

2.4.3 Cone and post insulators

The disc, cone, and post types of insulators are the most used shapes in GIS. Cone and Post shape insulators were studied and the effect of electric field on both were analysed. As illustrated in the figure below, a standard cone type and post type insulator is used for the current work. Distance between Conductor exterior surface and inner surface of the enclosure i.e. the clearance between the outer electrode to that of the inner electrode is 130 mm. While the outer electrode is grounded, a voltage of 1300kV is given to the inner electrode. SF6 gas is used to fill the gap, and 1.005 relative permittivity is assumed. The Cone Insulator is made of epoxy material. Electric field estimation is done by using FEM technique using software called electro 2d (version 9.2) It is described in more detail in the section below. Because rigid curves could enhance the field distributions seen in Fig. 2.1, the shape of the cone type spacer is adjusted considering the field spread acquired over the spacer surface. The 2D module is used to simulate insulator because of its symmetrical structure, and its electric field distribution. Below figure shows the equivalent circuit model of spacer.

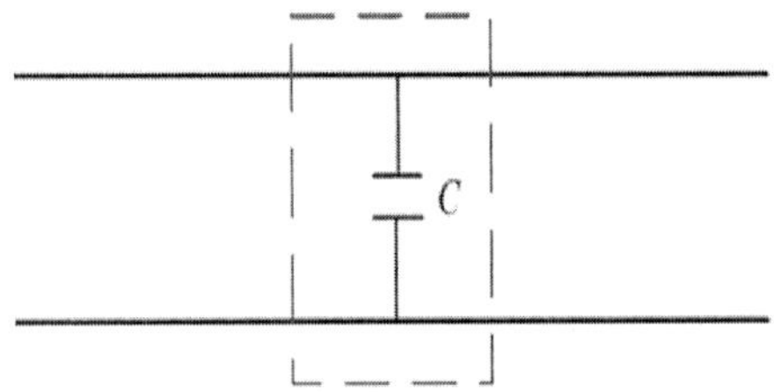

Fig 2.2: Equivalent circuit model of spacer

2.5 Electric field

A working knowledge of the field intensities in diverse media under electric shocks is crucial for high voltage engineers. Additionally, it aids in the selection of suitable electrode arrangements and cost-effective insulation sizing, preventing the formation of highly stressed regions and ensuring that the equipment operates dependably and lasts the expected amount of time. The field intensity E at any location in an electrostatic field is the ratio of the force on an infinitely small charge at that location to the charge itself as the charge decreases to zero.

The force F on any charge q is given by

$$F = qE$$

The electric flux density D is given by

$$D = \varepsilon E$$

Where ε is the permittivity of the medium in which the electric field exists.

The work done on a charge when moved in an electric field is defined as the potential. The potential is φ equal to

$$\varphi = -\int E\, dl$$

$$E = V/D$$

Where l is the path through which the charge is moved.

2.5.1 Electric field stresses

Like mechanical designs, where the criteria for design depend on the mechanical strength of the materials and the stresses produced during their operation, in high voltage applications, the dielectric strength of insulating materials and the electric field stresses developed in them when subjected to high voltages are the important factors in high voltage systems. Conductors and insulators are key components used in high voltage apparatuses. The insulators stop current from flowing in unwanted directions while the conductors transport the current. The insulation is the most crucial component of a high voltage device, as was already established. An insulating material's dielectric strength can be characterised as its ability to tolerate a maximum amount of dielectric stress. Pressure, temperature, humidity, field configurations, the type of applied voltage, flaws in the dielectric materials, the composition of the electrodes, and the surface of the test electrodes are only a few of the variables that affect the electric breakdown strength of insulating materials. Studying the potential methods through which the failure could occur will make it feasible to comprehend why the insulation failed.

The presence of discharges, either over the insulation's surface or inside its voids, is the most frequent reason for insulation failure. If such discharges could be eliminated at the typical working voltage, the likelihood of failure would be significantly lower. Failure can then happen because of the insulation degrading thermally or electrochemically.

2.5.2 Uniform and non-uniform electric fields

In general, the electric field between any two electrodes can be either uniform or non-uniform. In a uniform field gap, the average field E is constant across the field region, but in a non-uniform field gap, E varies at various locations inside the field region. When the gap distance is less than the sphere's diameter, uniform or nearly uniform field distributions exist between two infinite parallel plates or two spheres of equal diameter. For high voltage measurements for triggering in impulse voltage generation circuits, spherical electrodes are typically utilised. When the gap separation is significantly lower than the plate size, parallel plates of finite size are occasionally employed to imitate uniform electric fields. Most real-world high voltage parts used in electric power systems typically have irregular field distributions.

2.5.3 Applied voltage

The voltage applied or the potential difference across the insulator can also have an impact on the field stress that an insulator experience. While the applied voltage can be thought of as the energy per charge, the electric field can be thought of as the force that charged particles experience per unit of charge. The effort required to transport a unit test charge from one position to another is represented by the voltage difference between any two points, which can be written as V. As a result, the relationship between voltage difference and field can be summarised as follows:

$\vec{E} = - \Delta V \; \Delta d$ (14)

Where Δd represents the distance moved by the charge on applying a voltage and the negative sign indicates a decrease in potential in the field direction.

Basic lightning impulse insulation level

BIL stands for basic lightning impulse insulation level although this is often shortened to basic impulse level or basic insulation level. According to the records of the previous GIS systems, the 350 KV HVDC GIS the insulation design is based on BIL specified by the customer. However, the insulation design for 350 KV HVDC GIS require an impulse voltage of 1300KV according to the BIL Standard. BIL is not the same as the withstand strength of an insulation system. It is essentially the impulse level that an insulation system must withstand with a probability of failure below some value like 10% when subjected to a series of multiple impulses and each impulse is a standard shape with a specified rise time, crest value, decay rate and chopped tail or natural decay.

When power system equipment like Gas insulated switchgear are manufactured, the insulation design is based on a BIL specified by the customer and is related to the nominal system operating voltage. For higher voltage systems, multiple BIL levels are typically given in the standards to account for higher maximum operating voltages and where a system might be subjected to a higher-than-normal number of impulses.

2.6 Surface charges

High-Voltage-AC equipment is effectively employed in gas insulated substations (GIS). DC transmission lines are being erected and used at an increasing rate; however, air is still used as the isolation medium. It is necessary to consider extra issues to utilise the benefits of the GIS for DC-voltage. Surface charge accumulation creates an issue that lowers the system's breakdown voltage and causes a field distortion.

Additionally, unlike AC voltage, the conductivity k controls the electric field primarily in high voltage DC circumstances. The electrical setup of the spacer must be altered to increase field resolution. Due to the charges that have built up at the insulator interface, the electric field in an HVDC system is complicated and changes from being capacitive to resistive. The spacer's conductivity and surface charge are changed, and the effects of the field stress are examined.

2.7 Insulation properties of GIS insulators under lightning impulse with superimposed DC voltage

For instance, the potential distribution of GIS under dc voltage is determined by resistance properties, such as the electric charges that have built up on the insulator as a result of electrification. Numerous studies have been done on the resistance and electrification characteristics under dc voltage because these are essential elements for GIS dc insulation design [1, 12–16].

On the other hand, in the actual system, the range of electric stress on GIS encompasses both dc voltage and over voltages in varied time intervals. Although all of these over voltages serve as the foundation for insulation design, a lightning impulse (LI) overvoltage is the most important element. In the actual field, this LI overvoltage is typically superimposed on a dc working voltage for dc systems rather from being generated independently.

Therefore, it is essential to understand the insulation properties of GIS under mixed conditions with the LI overvoltage superimposed over the dc voltage (hereafter referred to as the "superimposed voltage"). In particular, insulators are affected by both the electrification condition and the electric field distribution under dc voltage, which could cause considerable changes in their insulation qualities under the superimposed voltage.

Chapter - 3

ELECTRO 2D Software

3.1 Introduction

To design and analysis of Support insulators for HVDC Gas insulated switchgear we use software called INTEGRATED ENGINEERING SOFTWARE 2D Electric Analysis Guide Version 9.2. This Software has the Physics, Solver, and Analysis menus for the 2D Electric Analysis module known as "ELECTRO". ELECTRO is designed to help solve the specific problems inherent in the design and analysis of electric equipment. ELECTRO incorporates a number of important features which work together to provide a fast, easy-to-learn, efficient, and reliable working tool.

Different physics modes are available in ELECTRO

These modes are:

1) Two-dimensional field solver delivers scalar and vector field distributions.
2) Rotationally symmetric field solvers deliver scalar and vector field distributions.

Three different analysis modes are incorporated into ELECTRO

The analysis modes are:

1) The static mode is for time-invariant electric fields.
2) The quasi-static phasor mode for time harmonic fields
3) The transient mode for general time varying fields

3.2 Assigning physical properties

To completely define a model, the geometry must have physical properties that define how the various parts of the geometry interact.

- Define the type of model (two-dimensional or rotationally symmetric).
- Define materials and add these materials to closed regions of the geometry.

- Assign boundary conditions to the interfaces.
- Assign sources, such as conductors, currents, and impressed fields.
- Define transient source.

3.3 Defining the type of model

There are five settings for the type of model.

- The geometry of a model represents two-dimensional or rotationally symmetric.
- The operation mode is static, phasor, or transient depending on the time dependence of the field.
- Define which material properties are relevant to the analysis.
- Define through charge balance (Rotational Symmetric only) how the voltages assigned relate to global ground.
- The setting is found from the Physics Global dialog, opened by selecting the menu **Physics > Physics Global Setup.**

3.4 Balanced and unbalanced rotational symmetric models

For some rotational symmetric models, you must force the total net charge to be zero. Such systems are "balanced". For balanced models, the voltage at infinity is between the minimum and maximum applied voltages of the system. Unbalanced is the correct choice for most Fields mode solutions, where the voltages specified are with respect to ground. You assume that the earth ground potential is zero volts. More formally, the ground is considered to be far away and is not part of the model. This implies that the voltage is zero at infinity. Balanced is normally the correct choice in Circuit Parameters mode since all charge must be accounted for.

3.5 Setting the material defaults

Define the type of materials used in the analysis. ELECTRO can solve static electric field problems.

For static problems, the constitutive equations are: $Jc = \sigma E$

Where E is the electric field intensity, Jc is the conduction current, σ is the constant called Electric Conductivity.

To set the material defaults, specify one of the following options in the **Physics Global Setup dialog:**

Permittivity- Perform a dielectric analysis that uses only the permittivity (dielectric constant) value of the materials in the model.

Conductivity- Perform a conductive analysis that uses only the conductivity value of the materials in the model.

Permittivity + Conductivity- Perform a lossy dielectric analysis that includes the permittivity and the conductivity values in the solution. This is only appropriate for Phasor and Transient solutions.

3.6 Adding materials

The Materials table is covered in the 2D Program Guide. Below are some items of particular concern to electric modeling.

The only Parameters the View/Edit Material dialog (the "Material Editor") which are relevant to ELECTRO are the Conductivity and Permittivity.

You can define linear materials, with a constant permittivity and conductivity.

3.7 Assigning boundary conditions

You can use boundary conditions to truncate models for faster computation time. The physics of a model determines the appropriate boundary condition that you should assign. The submenu Physics>Boundary Conditions contains commands for assigning, deleting and viewing boundary conditions.

Voltage (Uniform)

Assign a voltage boundary condition when you know the potential on part of the boundary (usually, conductors are connected to a voltage source or ground). Mathematically, this is equivalent to a Dirichlet boundary condition. To assign a uniform potential boundary condition:

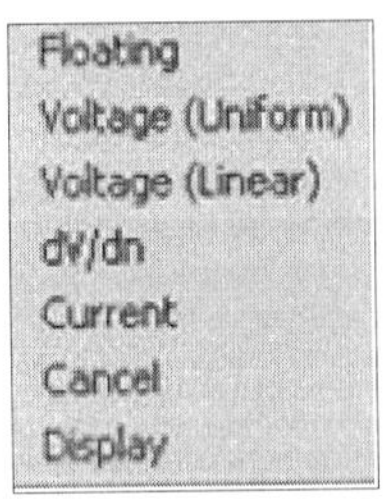

1) Select **Physics > Boundary conditions > Voltage (Uniform)**. In **Transient mode** select from the submenus for specific time dependencies.
2) Select the segments where you want to assign the potential.

3) In **Static mode**, enter the value of the potential that you want to assign. In **Transient mode** you will be prompted as appropriate for the specific time dependency selected.

4) Repeat steps 2) & 3) to assign to more geometry.

5) **Right-click** to exit when all assignments are made.

3.8 Assigning surface

You can assign surface to the geometry. The Surface Charge sub-menus of the Physics menu contain commands for assigning, deleting and displaying charges. These two menus contain four common commands: Assign Total, Delete, Display and Scale. The Surface Charge menu also contains an option to assign a non-uniform distribution specified in a file.

Assigning charges

To assign uniform charges:

- Select the **menu Physics > Surface charge > Assign total**
- Select the volumes (regions) or surfaces (segments) where you want to assign the charge.
- Enter the total charge that you want to assign.
- **Assign** more charges by repeating 2) & 3) or press the right mouse button to exit the Assign command.

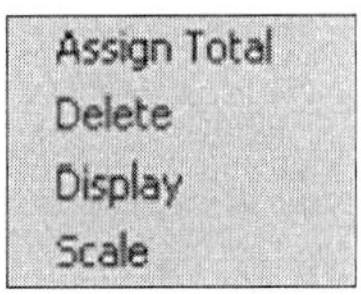

Editing charges

If you want to change an assigned charge, you can delete and re-assign the charge or you can scale the charge by a constant value. You can also display the charge assigned to a region.

To delete charges

- Select the menu **Physics > Surface charge > Delete**.
- Select the regions or segments where you want to delete the charge.
- Press the right mouse button to delete the charges and exit the **Delete** command.

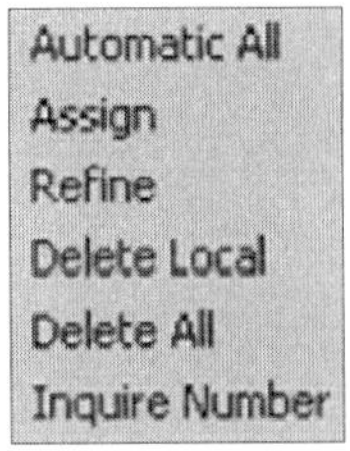

3.9 Generating 2D elements

3.9.1 2D elements and the BEM solver

When you assign volume sources or non-linear material to a region, in manual mode you must generate 2D elements in the region. ELECTRO's BEM solver uses 2D elements to distribute volume sources through a region and to calculate their effect on the field distribution.

The number of 2D elements in a charge-containing region has little effect on the solution accuracy as long as the subareas are not highly distorted. For regions with non-linear materials, increasing the number of 2D elements increases the accuracy of the field calculations. This is particularly true in regions where the fields vary rapidly within the material.

3.9.2 Boundary element method (BEM) in electric field

Although the charge simulation method is renowned for its speed and precision, it performs poorly when dealing with multi-dielectric issues and extremely thin electrodes, which are frequently used to regulate and measure the electric field strength in transformers, condenser bushings, and other devices. Surface Charge Simulation Method (SSM), also known as the Boundary Element Method (BEM), is a technique that can be used to address such issues. Cylinders, spheres, cones, and plane electrodes are frequently employed as electrodes and insulators in high voltage equipment. Because of this, it is important to consider elements with certain shapes to use variable charge distribution on these surfaces to simulate a field that is realistic.

To calculate complex geometries, the pieces are integrated in computer code. Only those positions where the *components* are inapplicable use curved triangles. This method generates a huge number of system equations since it requires a lot of elements-typically more than 2000-regardless of the geometry of the surface.

3.9.3 Principle of boundary element method

The Boundary Element Method's basic premise is that area elements are boundaries. BEM substitutes area charge elements for actual charges. The system components do not have to be symmetrical for BEM to apply, though. Real charges are often discretized using boundary elements with three or four nodes that apply linear shape functions to approximation the internal charge distribution. Suitable intermediary locations can be introduced if the geometry of the outer surface is not covered. There are BEMs that come in pre-configured shapes such spheres, cones, cylinders, and toroids. Numerical integration is used to evaluate the potential functions that the Boundary Element Method produces.

3.10 Creating graphs

You can plot any scalar field quantity along a line or geometry segment(s), or plot the fields tangential and normal to a geometry segment. Commands in Autograph allow you to integrate graphs and display information about graph points.

To create a graph

1) In the **View Type** selection box on the **Field Analysis Result** dialog, select the field quantity that you want to display. In the Display Form box, select Graphs.
2) In the **Location** of the graph. You can plot a graph along a line (straight line between two entered points) or along existing geometry segments.
3) Click **[New Graph].**
4) Enter the reference geometry for the graph. If you selected **Graph Along Line**, enter two points that define a line. For **Graph Along Segment**, select one or more geometry segments then **right-click.** Once you enter the reference points**, ELECTRO** calculates and displays the graph.

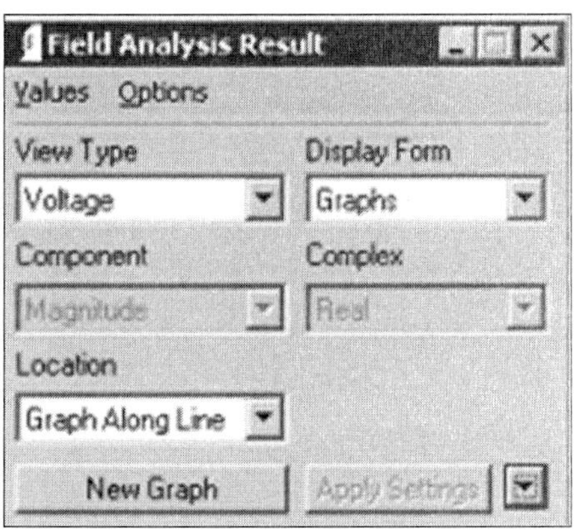

Once you create a graph, you can change the view settings options in the full dialog box and click **[Apply Settings]** to re-display the graph with the new options

Standard component signals

Step source

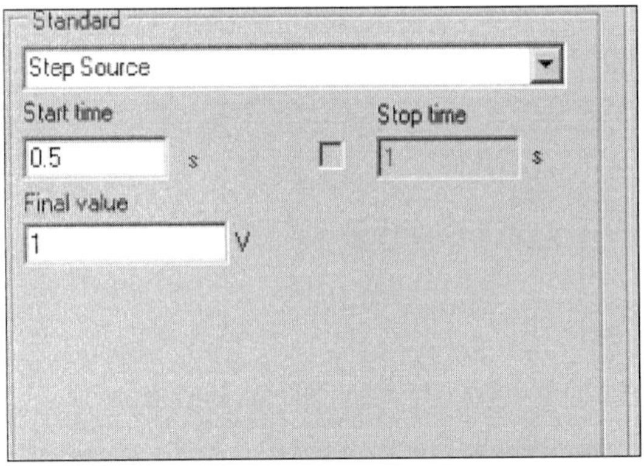

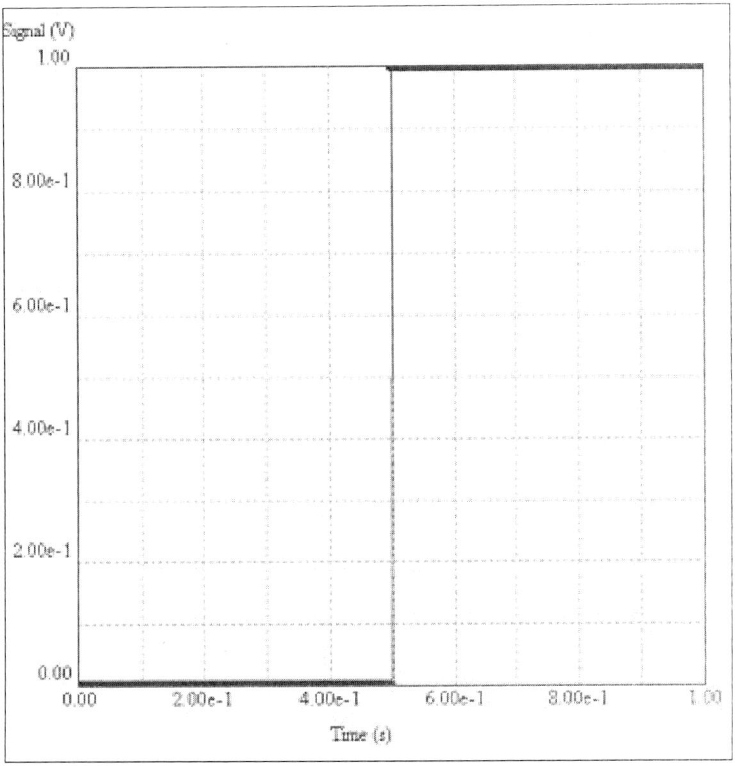
Signal (V)
1.00
8.00e-1
6.00e-1
4.00e-1
2.00e-1
0.00
0.00
2.00e-1
4.00e-1
6.00e-1
8.00e-1
1.00
Time (s)

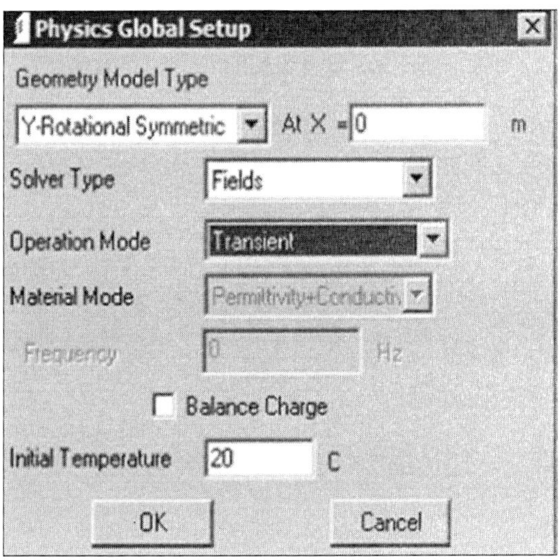
Physics Global Setup
Geometry Model Type
Y-Rotational Symmetric
At X = 0
m
Solver Type
Fields
Operation Mode
Transient
Material Mode
Permittivity+Conductiv
Frequency
0
Hz
Balance Charge
Initial Temperature
20
C
OK
Cancel

3.11 Static analysis

If the analysis is based on one instant in time and you calculate instantaneous field values and forces, or if you are looking for a "steady state" result a sufficient time after the sources have been applied, these are the Static cases.

3.12 Transient analysis

The most complex field solutions are those that require analysis of nonequilibrium or transient conditions. Solving these problems by numerical methods requires a series of field solutions at multiple discrete time instances. The **Physics Global Setup** dialog box has a transient Operation Mode setting as shown at right. When this setting is selected, the Transient section of the **Solver Setup** dialog is activated to select the **Transient Time Step Settings** for Total Time and number or steps (or step size).

Performing a transient analysis requires the following steps:

1) Create a geometric model and assign any required physical properties.
2) Set the Operation Mode to Transient.
3) Specify the types of time signals for all transient sources.
4) Specify the total time duration and the time interval between steps of the transient (Solver Setup dialog).
5) Run the analysis. Once the analysis is completed, post processing may be performed in the same manner as for any solved mode.

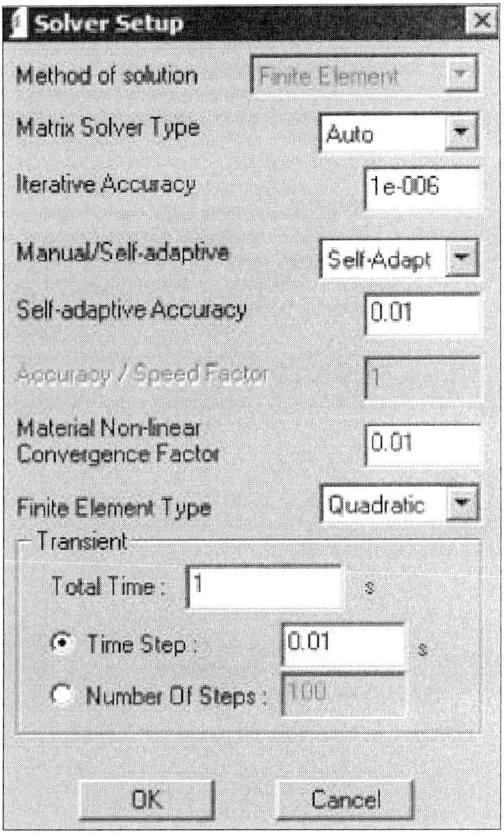

3.12.1 Setting the time variables in transient analysis

Numerical simulations of transients involve generating field solutions at multiple discrete time intervals. There are two-time parameters that must be specified in order to perform a transient analysis: the time step between solutions and the total duration of the transient.

When the Operation Mode of the Physics Global Setup dialog box is set for a transient analysis, the Transient section of the Solver Setup dialog is enabled.

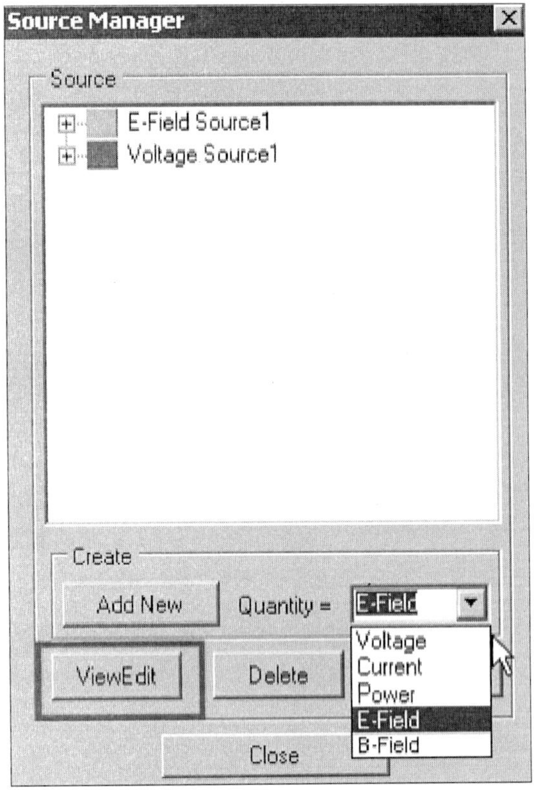

3.12.2 Types of transient source

ELECTRO transient sources can be either voltages or impressed electric fields. At right we've shown the Physics menu and emphasized the Boundary Conditions > Voltage and the Impressed E Field selections. Either of these commands will open the Source Table which is used to display and assign transient sources. You can also open the Source Table by selecting the last item of the Physics menu (this will only be active when the program is set for transient operation mode).

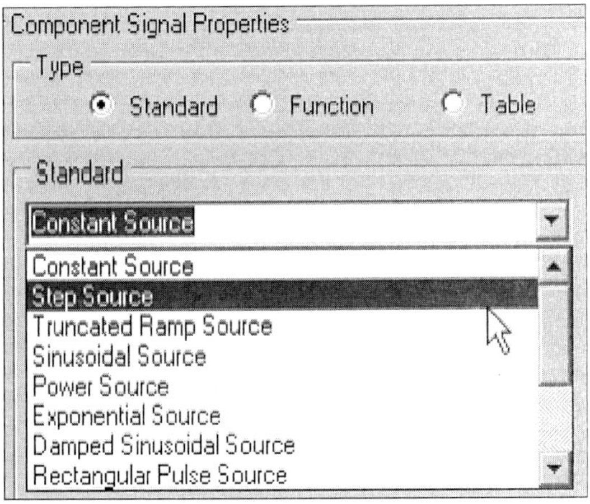

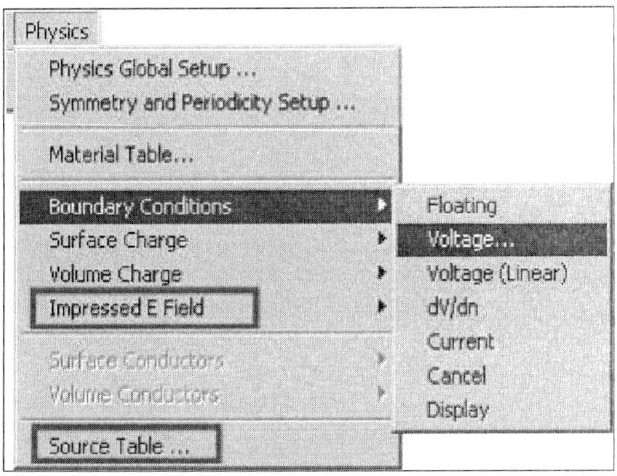

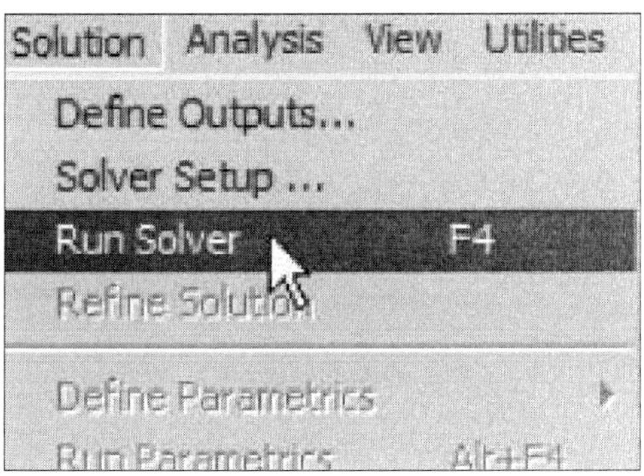

Initially, the Source Table will be empty as shown at right. Clicking the [Source Manager] button will open the Source Manager dialog box. The Source Manager (shown at right) allows you to add or delete Voltage or E-

Field sources (you can create other types of sources but the Electro field solver will only process Voltage or E-Field sources).

To examine or modify the time characteristics of a source, first select it in the Source Manager and then click the [View Edit] button to open the View/Edit Source dialog box.

3.12.3 Standard signal

You may be able to construct your desired source by using one or more of the Standard predefined time signals. At right we show examples of the Standard time signals available.

3.13 Running transient simulation

Transient simulations are run in exactly the same manner as other field solutions. Selecting Solution > Solve from the main program menu will initiate the solver. Transient simulations can only be run using the Finite Element solver. The program will automatically select this solver type and will not permit you to change to the Boundary Element solver while the Physics Global Setup dialog is set for transient analysis.

During the simulation the Run Solver dialog box will display information on the progress of the solution. When the analysis is completed, the Output Manager will automatically open if Outputs are defined. Otherwise, the Field Analysis Result dialog box will open.

3.14 Analysis of transient simulation results

Field plots and graphs of field components from transient simulations are produced using the Field Analysis Result dialog box, just as in static and time harmonic simulations. Since transient simulations are performed as multiple discrete time field solutions, you must set a specific time instant for plotting the field results. This is done in the At Time field as shown at right.

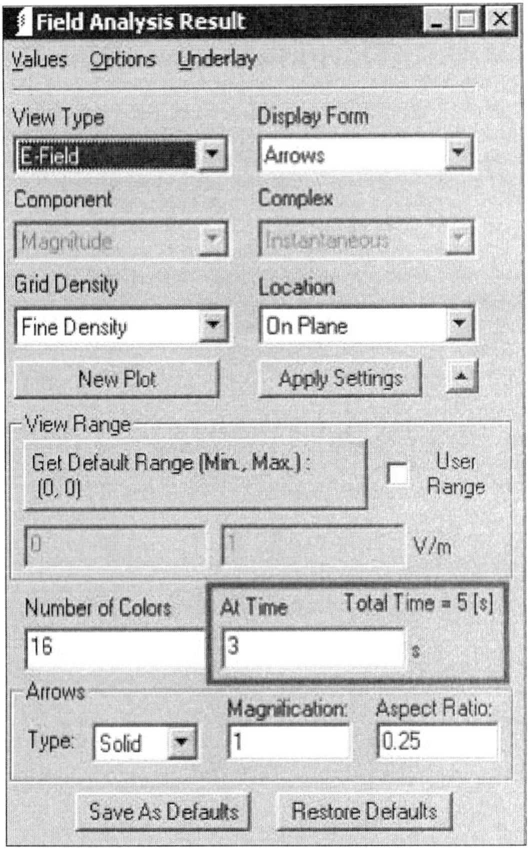
Field Analysis Result
Values
Options
Underlay
View Type
E-Field
Display Form
Arrows
Component
Magnitude
Complex
Instantaneous
Grid Density
Fine Density
Location
On Plane
New Plot
Apply Settings
View Range
Get Default Range (Min., Max.) : (0, 0)
User Range
0
1
V/m
Number of Colors
16
At Time
Total Time = 5 [s]
3
s
Arrows
Type:
Solid
Magnification:
1
Aspect Ratio:
0.25
Save As Defaults
Restore Defaults

Chapter - 4

Electrostatic Field Analysis using Electro 2D

4.1 Introduction

The electrostatic field analysis was carried out for the following cases: a) field analysis without insulator, b) post insulator and c) cone insulator. These configurations were analysed for static and transient conditions to know the maximum of electric field in each case. To analyse the field effect first case with no insulator was considered. Then a Post Insulator with minimum insulation distance between high voltage and ground was considered to notice the effect, subsequently the optimised stress i.e., geometric profile (Cone Insulator) generally used in GIS was analysed.

To carry out the HVDC analysis, transient and static analysis are performed. In static analysis the operation mode is only with permittivity and only one instant the results are observed, it is time invariant analysis. In transient analysis the operation starts with initially permittivity of the material and later conductivity of the material come into picture.

4.2 Simulation

Gas insulated switchgear (GIS) for high voltage direct current (HVDC) requires solid insulating materials or insulators to separate pressurized gas compartments and support wires mechanically. The presence of the insulator has a negative impact on the breakdown of SF6 gas insulation, particularly where the solid insulating spacer and SF6 gas meet. Because of its extremely high dielectric strength, SF6 is a 100 times more effective interrupting medium than air. The SF6 breakdown strength is 9 KV/mm.

Hence there is a need for control of electric field levels on these insulators in order to reduce its internal and surface discharges. Most of the failures in GIS system are due to improper design of insulator both in terms of profile and dimensions. Thus, their shape plays an important role in withstanding electrostatic field levels.

4.3 Analysis of cone and post insulator

The electrostatic field analysis was carried out for the following cases:

a) field analysis without insulator, b) post insulator and c) cone insulator. These configurations were analysed for static and transient conditions to know the maximum of electric field in each case. To analyse the field effect first case with no insulator was considered. Then a Post Insulator with minimum insulation distance between high voltage and ground was considered to notice the effect, subsequently the optimised stress i.e., geometric profile (Cone Insulator) generally used in GIS was analysed. From the comparison of NO insulator, static and transient analysis, we noticed thar cone insulator has less stress compare to post and without insulator. So, we varied the conductivity of cone insulator from $6.6e^{-10}$ S/m to $6.6e^{-22}$ S/m. From the result we analysed that for the conductivity of $6.6e^{-14}$ S/m has multiple variation of stress, so we considered $6.6e^{-14}$ S/m for the analysis of all geometrical profiles.

To carry out the HVDC analysis, transient and static analysis are performed. In static analysis the operation mode is only with permittivity and only one instant the results are observed, it is time invariant analysis. In transient analysis the operation starts with initially permittivity of the material and later conductivity of the material come into picture.

4.4 Steady state analysis

The AC analysis is also called as static analysis and this mode is for time-invariant electric fields. If the analysis is based on one instant in time and you calculate instantaneous field values and forces, or if you are looking for a "steady state" result a sufficient time after the sources have been applied, these are the Static cases. The simplest types of field solutions are those obtained by assuming equilibrium or steady state conditions produced by constant time-invariant sources. When these conditions are met, the Operation Mode of the Physics Setup Dialog should be set to perform a Static analysis.

In static analysis electric field distribution across insulated spacer under DC voltage condition is determined by the permittivity of the insulation material. Static analysis is based on one instant of time, Voltage distribution and Electric field stress distribution is calculated for post and cone insulator using static analysis.

4.5 Transient analysis

While receiving a lightning impulse voltage, gas-insulated switch-gears (GIS) experience a transitory electric field. The Transient option solves the model repeatedly throughout time as opposed to the Static and Phasor choices, which only require a single solution. When a boundary is excited, such as by applying a polarity reversal voltage to the insulation of an HVDC device, permittivity and resistivity in the insulating media change from one steady-state to another, creating a transient electric field. The ratio of the epoxy material's and SF6 gas's permittivity (also known as the capacitive distribution) controls the electric field distribution throughout the insulator and the area around it during the initial period of voltage application. The conductivity of the epoxy and SF6 gas then determines the field distribution (resistive distribution). For a specific insulator arrangement, the time constant and transition of the electric field from capacitive to resistive are computed. The study includes an analysis of the impact of insulating material conductivity on field dispersion. It is also addressed how the electric field changes along the insulator surface over time following the application of DC voltage.

Different conductivity values, ranging from 6.6e-8 S/m to 6.6e-22 S/m, were taken into consideration to explore the impact of insulator conductivity during the transient analysis. For each instance, the field effect was examined. Each example was given a time limit of 20,000 seconds. The greatest change in the field over time was seen at conductivity values of 6.6e-12 S/m and 6.6e-14 S/m. Only conductivities of 6.6e-12 S/m and 6.6e-14 S/m were included in the electric field stress analysis of the POST insulator. The maximum electric field stress, which was determined from the field data, was examined in order to maximise the electric field stress.

4.6 Methodology

2D model creation in Electro: 2. Defining Material & Boundary Condition 3. Defining Element & Simulation Run- (Number of elements, Conductivity & time period) 4. Post processing (Graph generation and Results)

4.6.1 C2D model creation in Electro

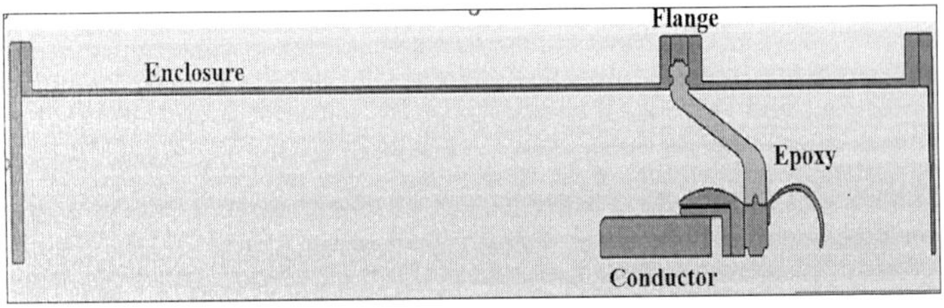

Fig 4.1: Cone epoxy spacer using HVDC analysis

The GIS has a stainless steel or aluminum sealed metal casing that maintains the gas under pressure and prevents SF6 leakage. Due to the tiny dimensions, it can be positioned outside or within a building. Although steel is also frequently used, aluminum is typically cast or welded for enclosures. In the centre of a metal tubular container are high voltage conductors. Insulator provides support for the conductor. Pure aluminum is used to make conductors, and a high potential is applied to the conductor to facilitate the passage of electrical energy.

Shield is designed to reduce localized stress on any path and to improve field distribution. Shield has a copper-chromium finger contact that is utilized to transfer current.

Aluminum used in the flange has zero potential voltage. GIS enclosure modules are attached to one another using bolted flanges that seal the enclosure with "O" rings. The GIS modules' many metal enclosure pieces are electrically connected to one another. The effectiveness of the GIS's electrical insulation is most significantly impacted by the epoxy spacer, which serves as a supporting conductor and electrical insulator. The epoxy resin used to create these support insulators is carefully cast to prevent the development of voids or cracks during the curing process.

In this study, a cone insulator's existing design, which is employed in HVAC 420KV GIS, is applied to HVDC 350KV GIS. According to the Basic Lightning Impulse Insulation Level (BIL) Standard, this insulation design for 350 KV HVDC GIS must be able to withstand an impulse voltage of 1300KV. It is essentially the maximum amount of impulses that an insulation system can withstand with a likelihood of failure lower than a certain threshold, say 10%, when each impulse has a set form and rise time.

4.6.2 Defining material and boundary condition

Geometry was created as shown in the above figure and uniform boundary conditions was applied. Material is defined for enclosure, conductor and insulator as shown in the figure. The material used for enclosure and conductor is pure aluminium and for insulator is standard molded epoxy. Uniform voltage was applied for conductor and enclosure, for conductor high potential 1300 KV is applied and for enclosure zero voltage is applied.

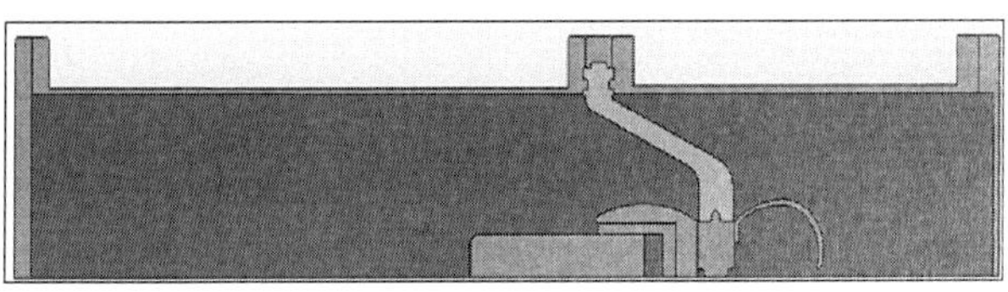

Fig 4.2: Assigning material properties to the cone insulator

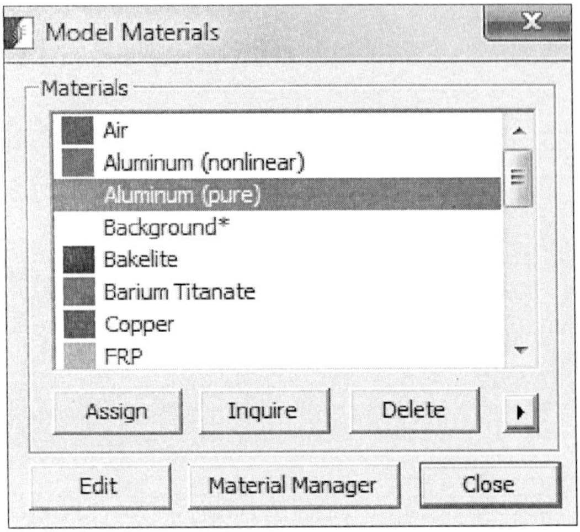
Model Materials
Materials
Air
Aluminum (nonlinear)
Aluminum (pure)
Background*
Bakelite
Barium Titanate
Copper
FRP
Assign
Inquire
Delete
Edit
Material Manager
Close

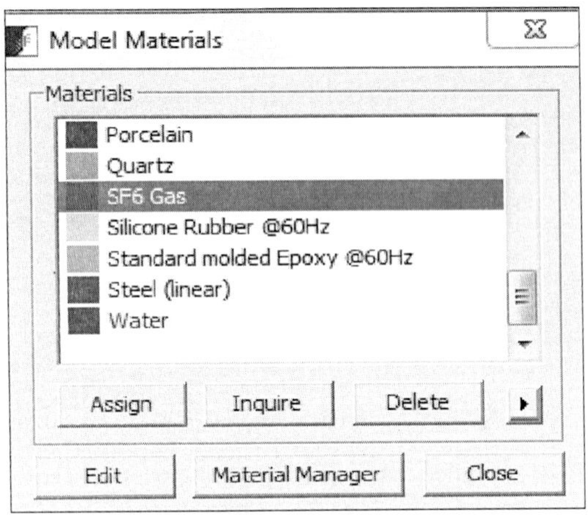
Model Materials
Materials
Porcelain
Quartz
SF6 Gas
Silicone Rubber @60Hz
Standard molded Epoxy @60Hz
Steel (linear)
Water
Assign
Inquire
Delete
Edit
Material Manager
Close

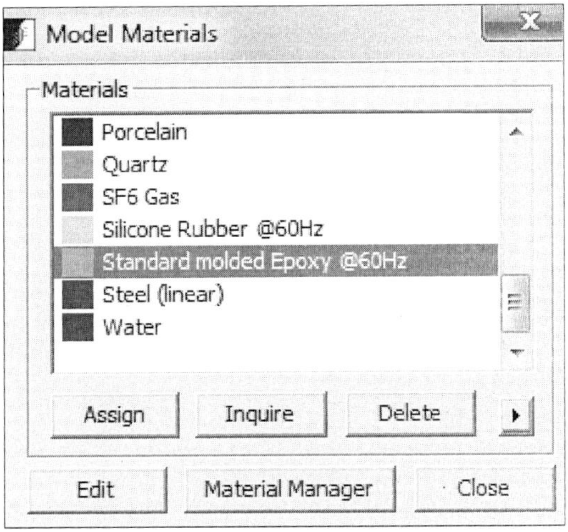

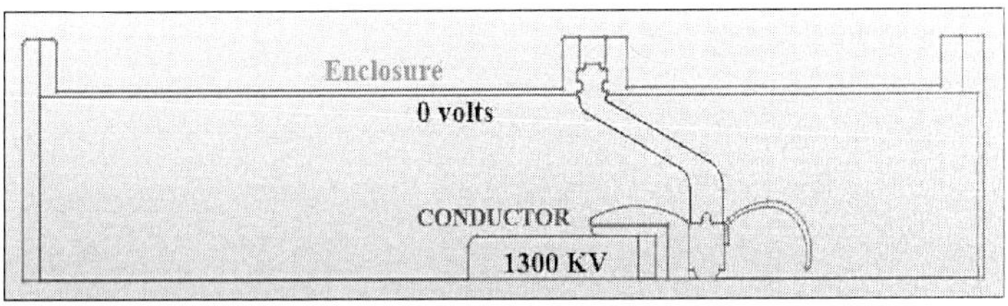

Fig 4.3: Assigning uniform voltage boundary conditions to the dielectric region

Table 4.1: Representation of material, Boundary voltages and default elements values.

Conductor	Material used: Aluminium	Applied voltage: 1300 kv	Conductivity: 36900000 s/m
Enclosure	Material used: Aluminium	Applied voltage: 0 v	Conductivity: 36900000 s/m
Epoxy	Material used: Standard moulded epoxy	-------	Conductivity: $6.6e^{-18}$ s/m

4.6.3 Defining element and simulation run

The electric conductivity of aluminium considered for present study is 36900000 S/m and conductivity of Epoxy as $6.6e^{-14}$ S/m (Default value of conductivity of Epoxy is $6.6e^{-18}$ S/m) The conductivity of aluminium components is constant and only the conductivity values for Epoxy was varied to study the effect of various conductivities on the insulator over a period of time. Time period of 5.5hrs i.e., 20,000 sec to check the field Analysis Result graph along segment for the time period. After assigning all the required values the Simulation was run and results were studied.

Assigning electric conductivity for aluminium and epoxy

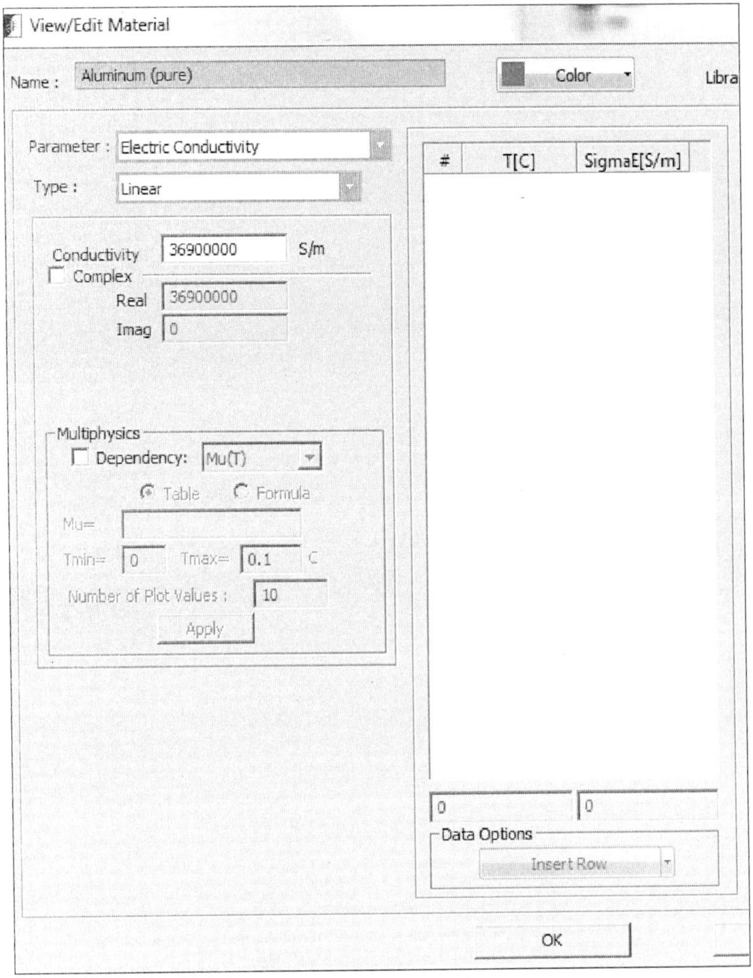

Fig 4.4: Assigning electric conductivity for aluminium

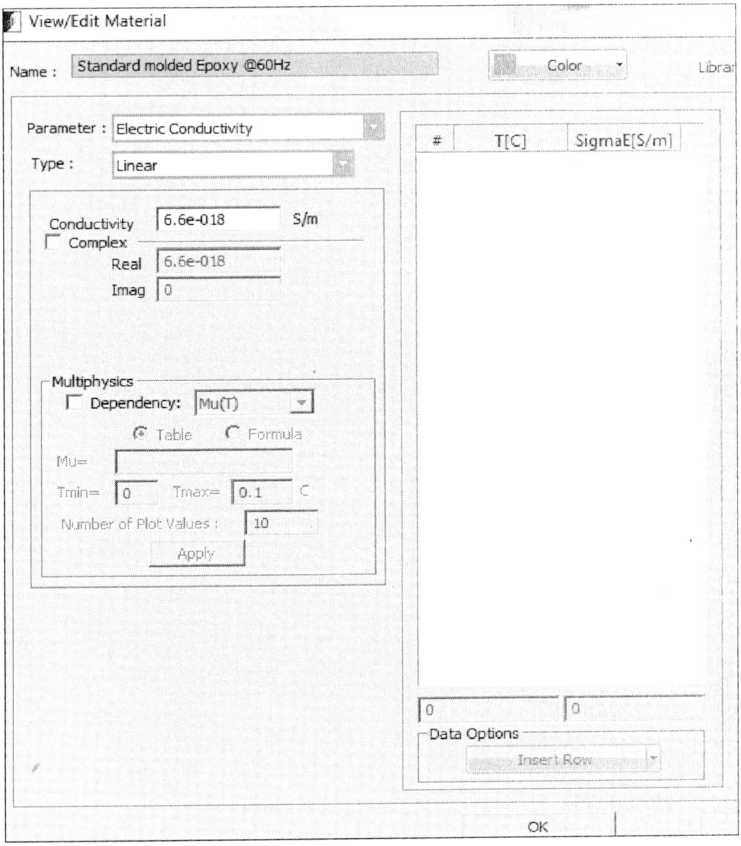

Fig 4.5: Assigning of electric conductivity for epoxy

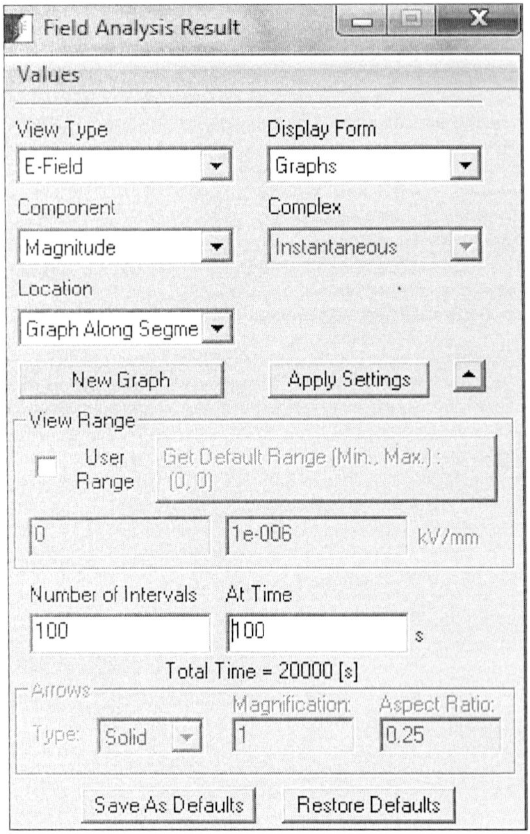

Fig 4.6: Field analysis result

Chapter - 5

Simulation Results

5.1 Results of static analysis

5.1.1 Case 1: V and E distribution with NO insulator using static analysis

The static analysis is for time-invariant electric fields. The below figure shows the voltage distribution of HVDC GIS with NO insulator using static analysis. The electrostatic field stress between HT conductor and LT enclosure at different positions is presented in the figure below.

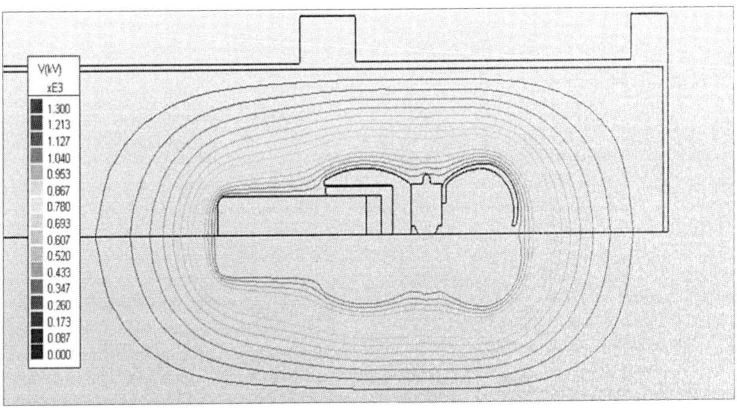

Fig 5.1: Voltage distribution of HVDC GIS with NO insulator using static analysis

From the below results of field distribution with NO insulator we can observe that with no insulator the electric field stress is approximately 25 kV/mm.

Position 1: The electric stress values between HT conductor to LT enclosure = 25.5 KV/m.

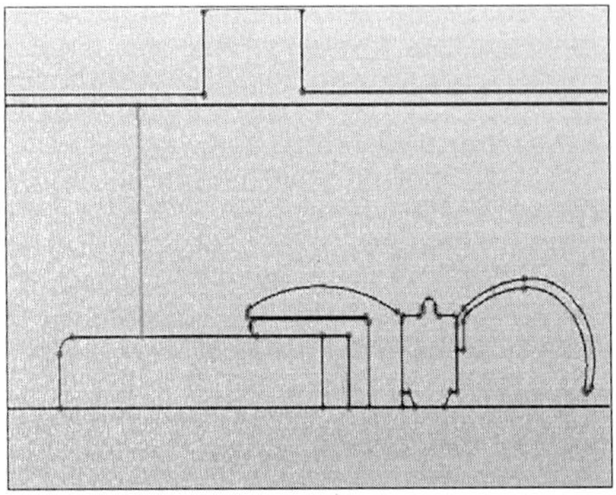

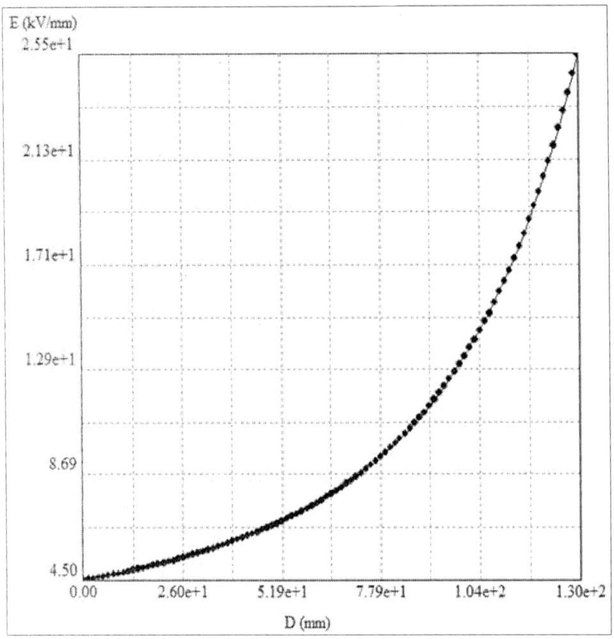

Position 2: The electric stress values between HT conductor to LT enclosure is 25 KV/mm.

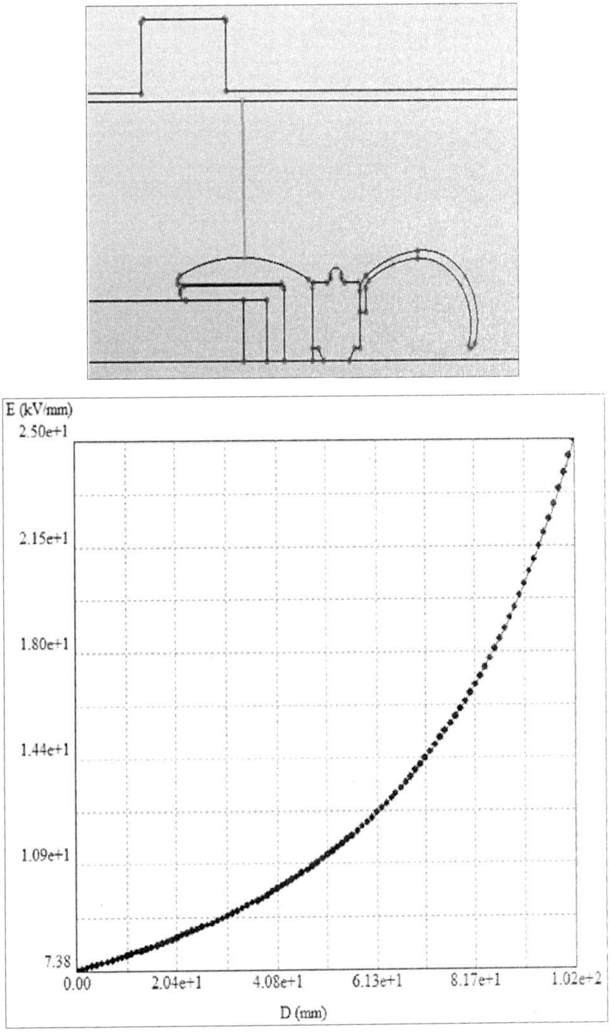

Fig 5.2: Results of E-field at Position-1 and Position-2

5.1.2 Case 2: V and E distribution with post insulator using static analysis

The below figure shows the voltage distribution of HVDC GIS with post insulator using static analysis. From the result we can observe voltage distribution and electro static field stress of post insulator for the conductivity of $6.6e^{-14}$ S/m.

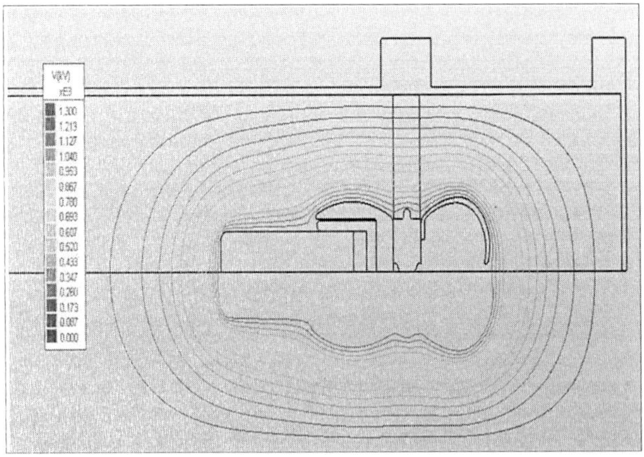

Fig 5.3: Voltage distribution of HVDC GIS with post insulator using static analysis

Voltage distribution across post insulator = 1303 KV

Electrostatic field stress for concave and convex side of post insulator = 16.7 KV/mm.

We can observe that as it is post insulator the concave side and convex side has same value of electric field.

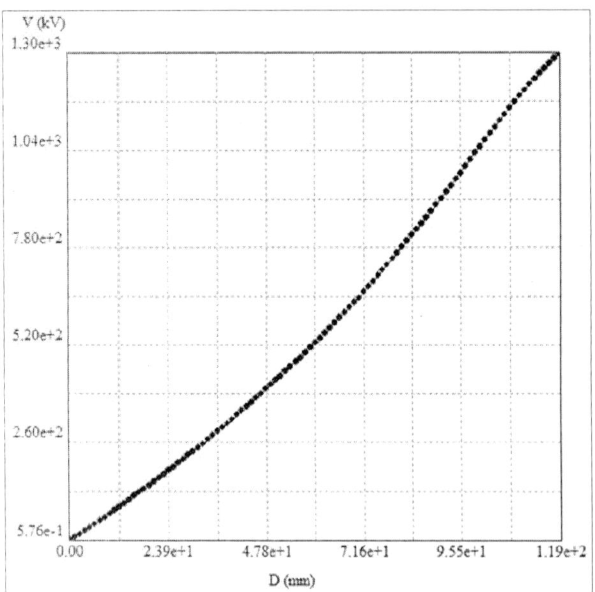

Voltage profile

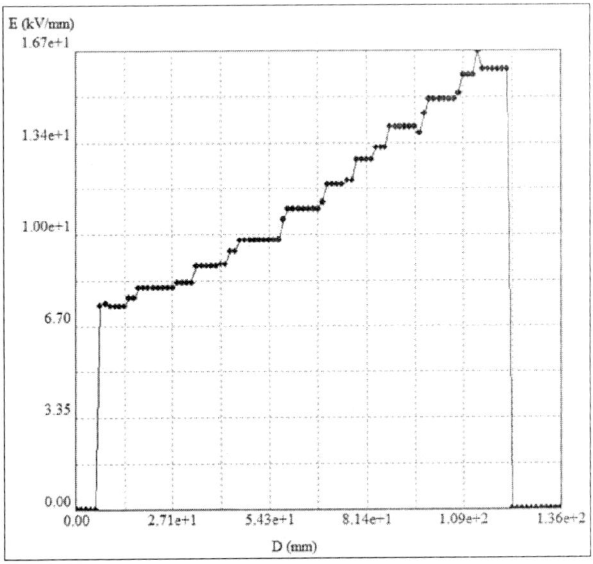

E-field result

Fig 5.4: Voltage profile and E-field result

5.1.3 Case 3: V and E distribution with cone insulator using AC analysis

The below figure shows the voltage distribution of HVDC GIS with cone insulator using static analysis. From the result we can observe voltage distribution and electro static field stress of cone insulator for the conductivity of $6.6e^{-14}$ S/m.

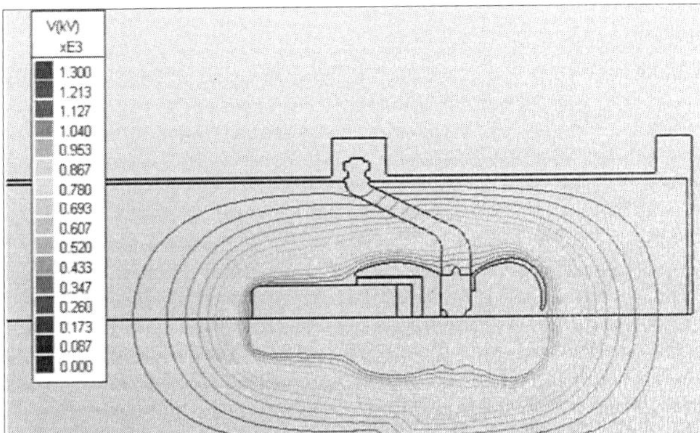

Fig 5.5: Voltage distribution of HVDC GIS with cone insulator using static analysis

Electric field stress on concave side of cone insulator = 14.1 KV/mm

Electric field stress on convex side of cone insulator = 20. 7 KV/mm

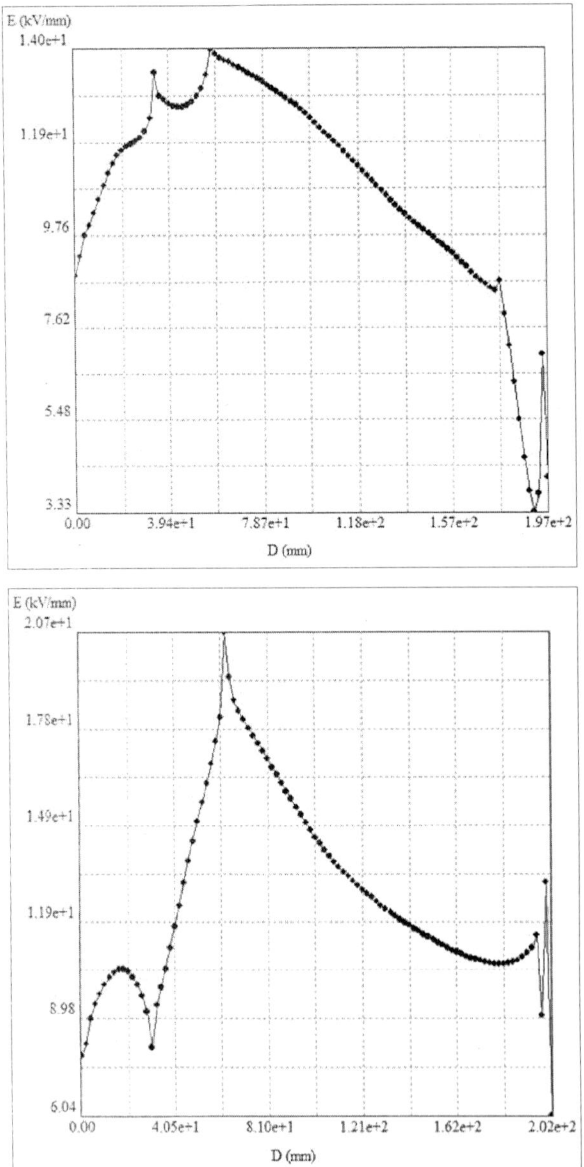

Fig 5.6: E-field result of concave & convex side of cone insulator

5.2 Results of transient analysis

5.2.1 Case 1: E-field distribution with NO insulator using transient Analysis

The transient analysis used for time varying fields. Below figure shows the design model of HVDC GIS with NO insulator using transient analysis. Calculate the electrostatic field stress between HT conductor and LT enclosure at different positions.

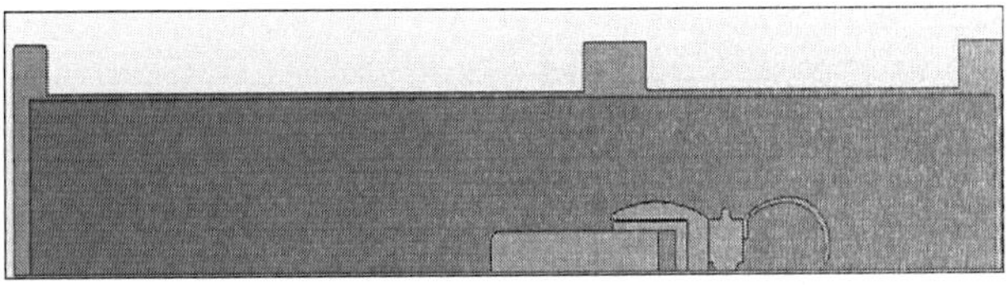

Fig 5.7: Design model of HVDC GIS with NO insulator using transient analysis

Position 1: The electric stress values between HT conductor to LT enclosure is 17.2 KV/mm

Figure shows the variation of electric field stress (KV/mm) with respect to distance (mm). From the results we can observe that Electric field stress with no insulator the electric field stress is approximately 17 KV/mm.

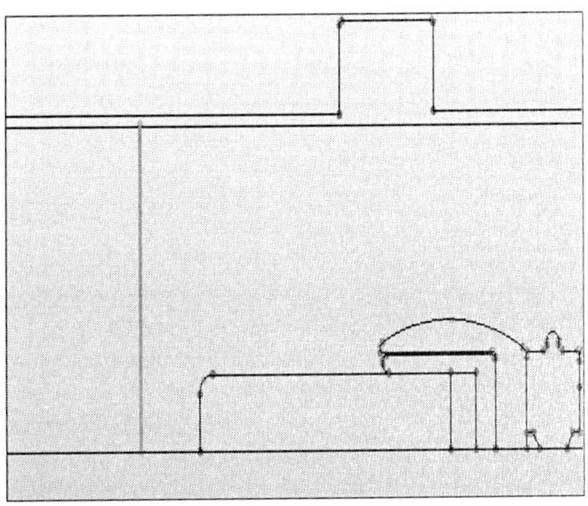

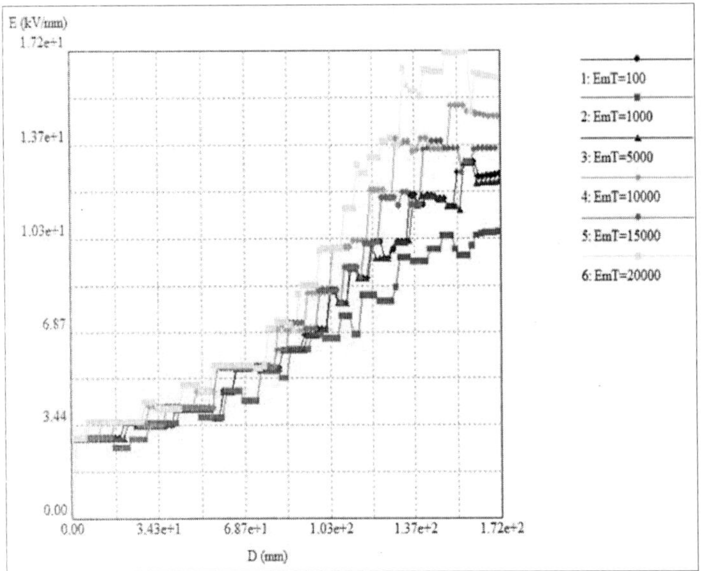

Representation of electric field stress vs. Time

Figure shows the variation of electric field stress (KV/mm) with respect to time (mm).

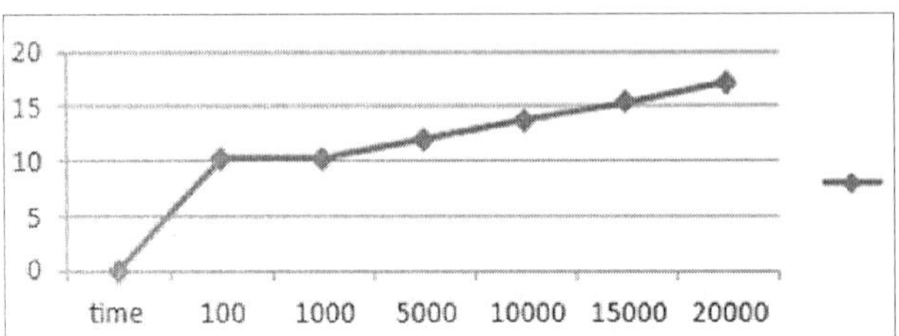

Fig 5.8: E-field result for position-1 and plot between Time vs Stress

Table 5.1

Time	Stress
100	1.03E+01
1000	1.03E+01
5000	1.20E+01
10000	1.37E+01
15000	1.54E+01
20000	1.72E+01

Position 2: The electric stress values between HT conductor to LT enclosure is 37.2 KV/mm

Figure shows the variation of electric field stress (KV/mm) with respect to distance (mm). From the results we can observe that with no insulator the electric field stress is approximately 37 KV/mm.

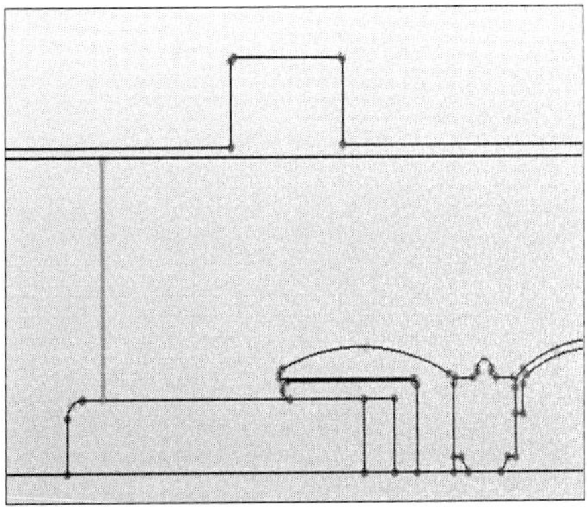

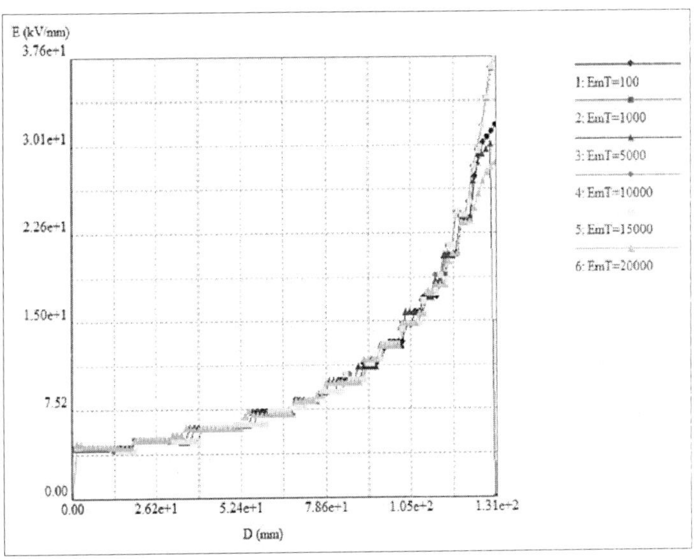

Representation of electric field stress vs time

Figure shows the variation of electric field stress (KV/mm) with respect to time (sec).

Table 5.2

Time (Sec)	Stress (KV/mm)
100	3.01E+01
1000	3.76E+01
5000	3.76E+01
10000	3.76E+01
15000	3.76E+01
20000	3.76E+01

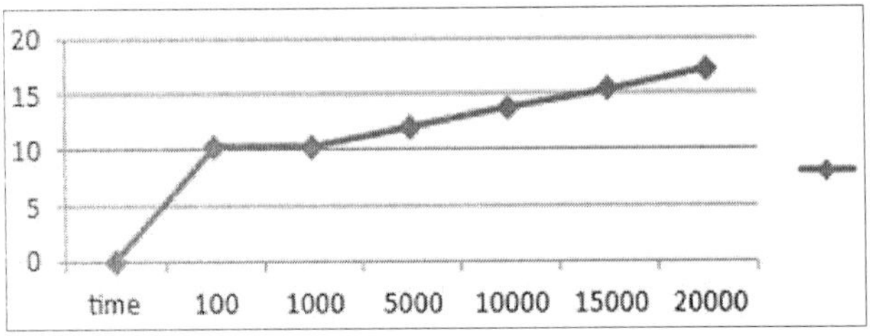

Fig 5.9: E-field result for position-2 and plot between time vs stress

Position 3: The electric stress values between HT conductor to LT enclosure is 23.6 KV/mm

Figure shows the variation of electric field stress (KV/mm) with respect to distance (mm). From the results we can observe that with no insulator the electric field stress is approximately 24 KV/mm.

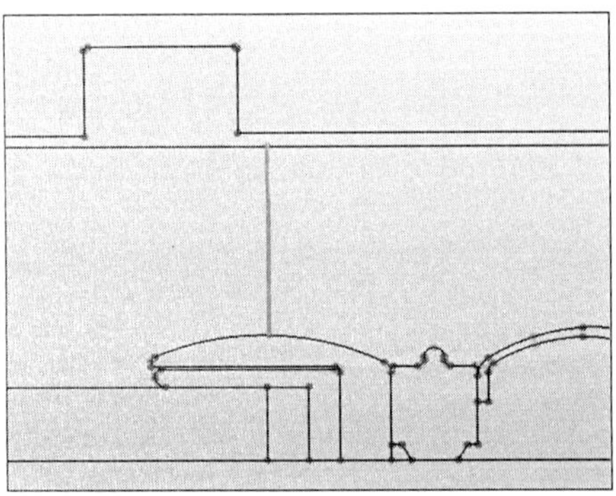

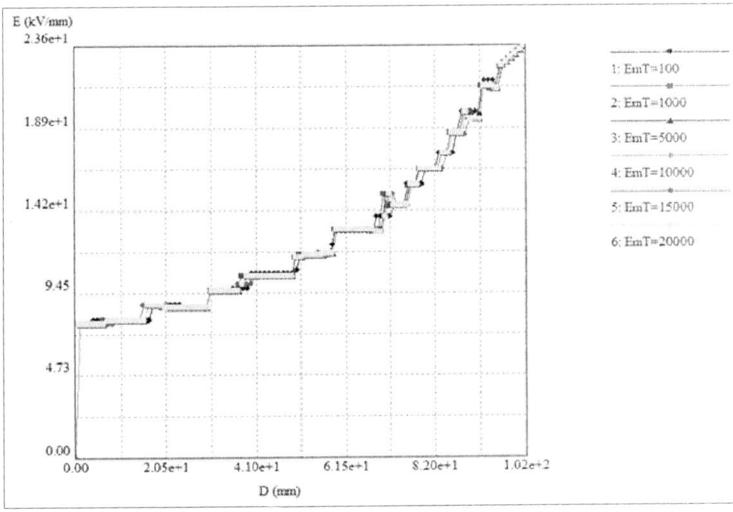

Representation of electric field stress vs time

Figure shows the variation of electric field stress with respect to time.

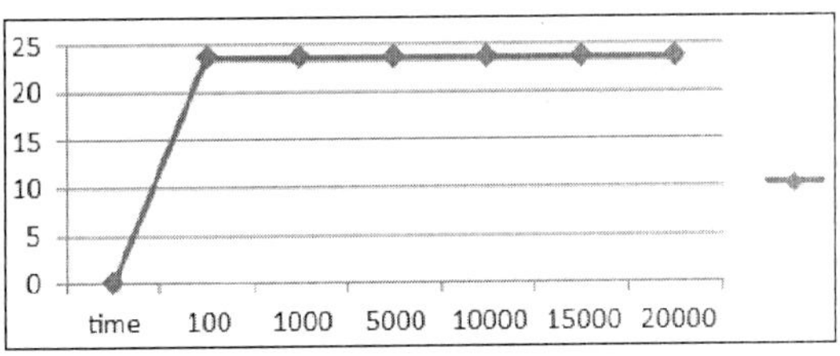

Fig 5.10: E-field result for position-3 and plot between Time vs Stress

Table 5.3

Time	Stress
100	2.36E+01
1000	2.36E+01
5000	2.36E+01
10000	2.36E+01
15000	2.36E+01
20000	2.36E+01

5.2.2 Case 2: E-field distribution with post-insulator using transient analysis

The transient analysis used for time varying fields. Below figure shows the design model of HVDC GIS with post insulator for the conductivity of

6.6e^{-14}S/m. Calculate the electric field stress on concave and convex side of post insulator.

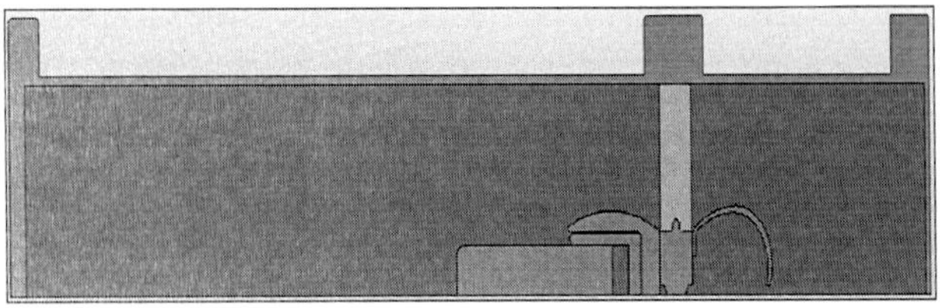

Fig 5.11: Design model of HVDC GIS with post insulator using transient analysis

Concave side of post insulator

Figure shows the variation of electric field stress (KV/mm) with respect to distance (mm) of the concave side post insulator. Maximum stress obtained is 15.4 KV/mm.

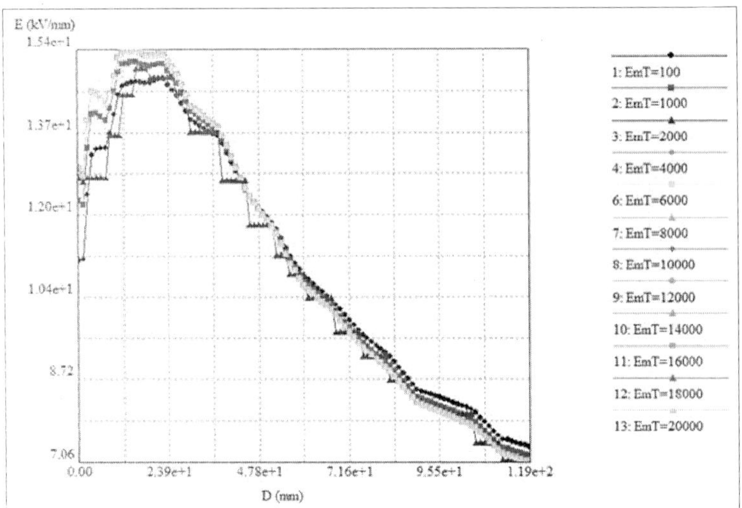

Fig 5.12: E-Field result of concave side of post insulator

Representation of electric field stress vs time for concave side of post insulator

Below figure shows the variation of electric field stress with respect to time.

Table 5.4

Time	Stress
100	1.46E+01
1000	1.51E+01
2000	1.51E+01
4000	1.54E+01
6000	1.54E+01
8000	1.54E+01
10000	1.54E+01
12000	1.54E+01
14000	1.54E+01
16000	1.54E+01
18000	1.54E+01
20000	1.54E+01

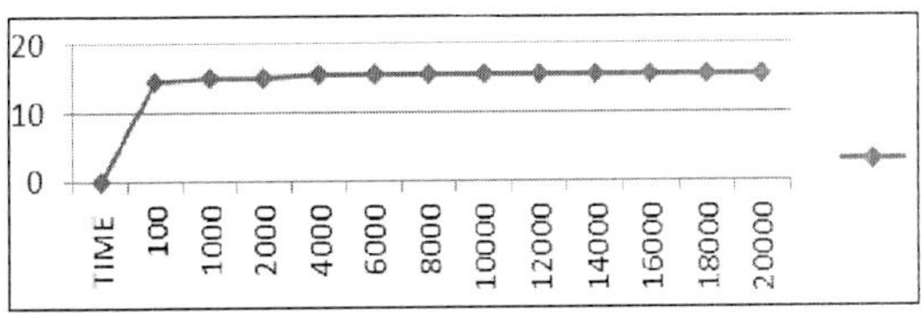

Fig 5.13: Plot between Time vs Stress on concave side of post insulator

Convex side of post insulator

Below figure shows the variation of electric field stress with respect to distance of the convex side of post insulator.

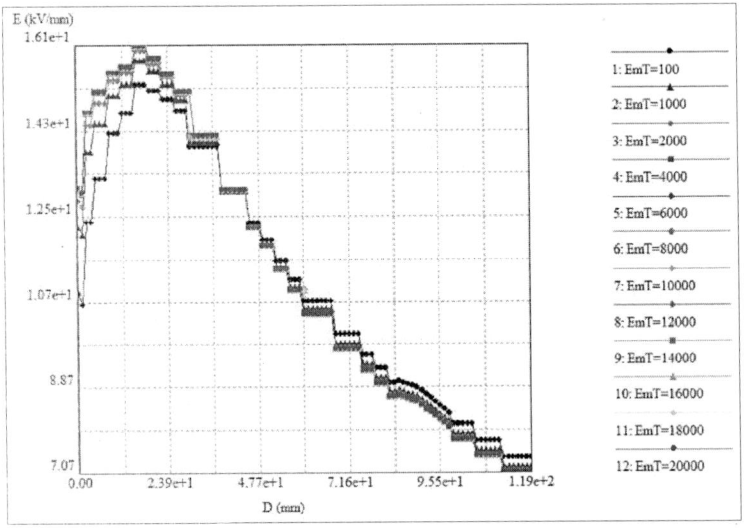

Fig 5.14: E-field result of convex side of post insulator

Representation of Electric field stress vs Time for convex side of post insulator

Table-5.5

Time	Stress
100	1.46E+01
1000	1.51E+01
2000	1.51E+01
4000	1.54E+01
6000	1.54E+01
8000	1.54E+01
10000	1.54E+01
12000	1.54E+01
14000	1.54E+01
16000	1.54E+01
18000	1.54E+01
20000	1.54E+01

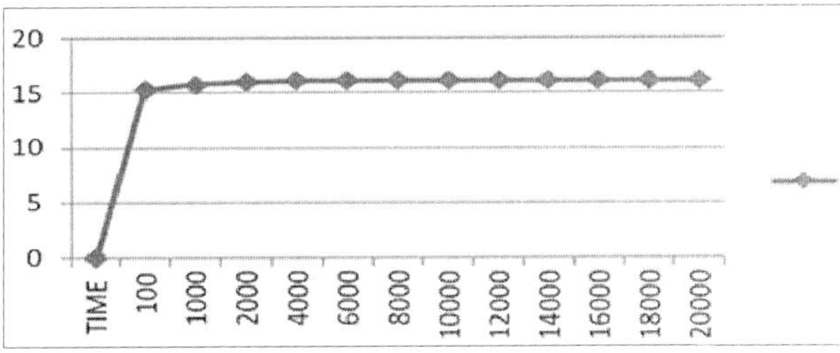

Fig 5.15. Plot between Time vs Stress on convex side of post insulator

5.2.3 Case 3: e-field distribution with cone insulator using transient analysis

To study the effect of insulator conductivity during the transient analysis various values of conductivity was considered from $6.6e^{-10}$ S/m to $6.6e^{-22}$ S/m. The field effect was studied for each case. The time duration for each case was 20000 sec. For a conductivity value of 6.6e-12 S/m and 6.6e-14 S/m, maximum variation in the field over time was observed. Electric Field Stress analysis for POST insulator was carried out for conductivity of $6.6e^{-12}$ S/m and $6.6e^{-14}$ S/m only. Default value of conductivity of cone epoxy insulator is $6.6e^{-18}$ S/m.

The transient analysis used for time varying fields. Below figure shows the design model of HVDC GIS with cone insulator for the conductivity of $6.6e^{-14}$ S/m. Calculate the electric field stress on concave and convex side of cone insulator.

Fig 5.16: Geometric model of HVDC GIS with cone insulator using transient analysis

Concave side of epoxy

From the below figure, the Maximum Electric Stress obtained is 13.9 KV/mm.

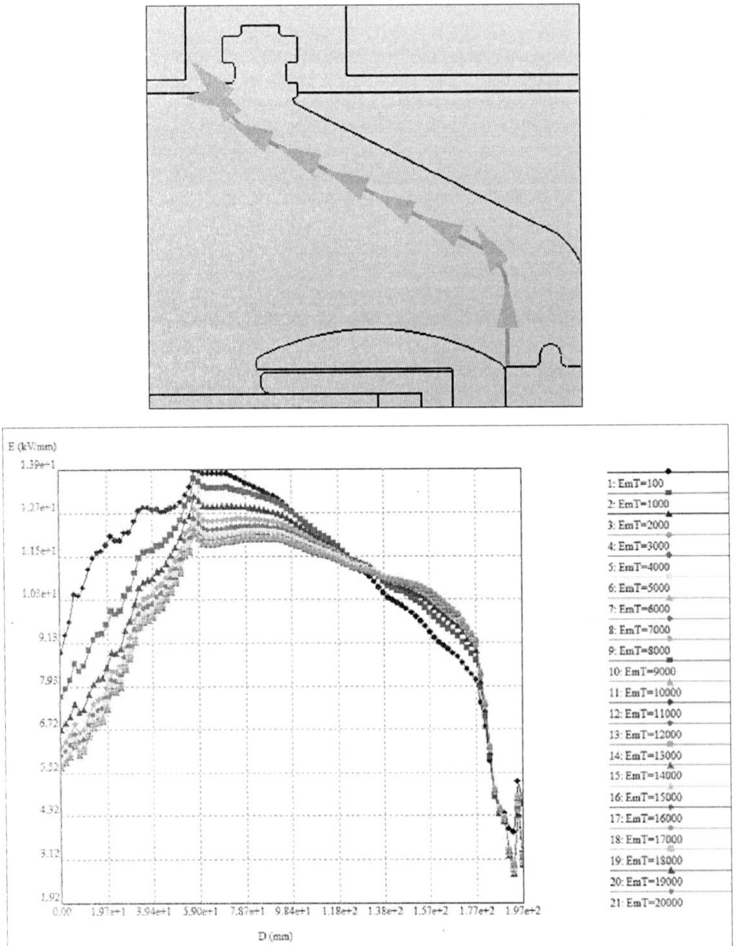

Fig 5.17: E-field result of concave side of cone insulator

Representation of electric field stress vs time of concave side of cone insulator

Table 5.6

Time (sec)	Stress (KV/mm)
100	1.39E+01
500	1.39E+01
1000	1.37E+01
2000	1.32E+01

3000	1.28E+01
4000	1.26E+01
5000	1.25E+01
6000	1.25E+01
7000	1.25E+01
8000	1.25E+01
9000	1.25E+01
10000	1.25E+01
11000	1.25E+01
12000	1.25E+01
13000	1.25E+01
14000	1.25E+01
15000	1.25E+01
16000	1.25E+01
17000	1.25E+01
18000	1.25E+01
19000	1.25E+01
20000	1.25E+01

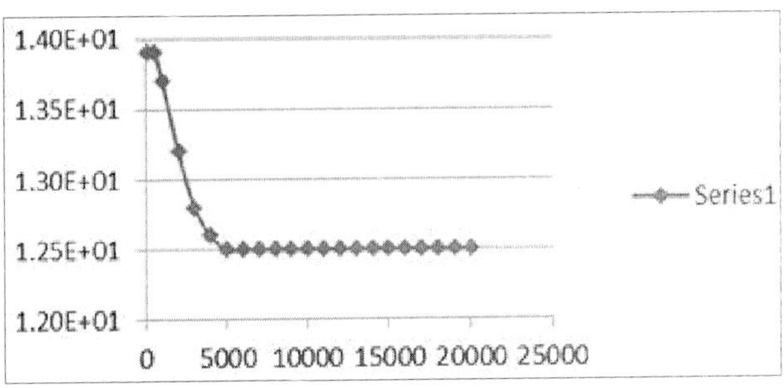

Fig 5.18: Plot between Time vs Stress for concave side of cone insulator

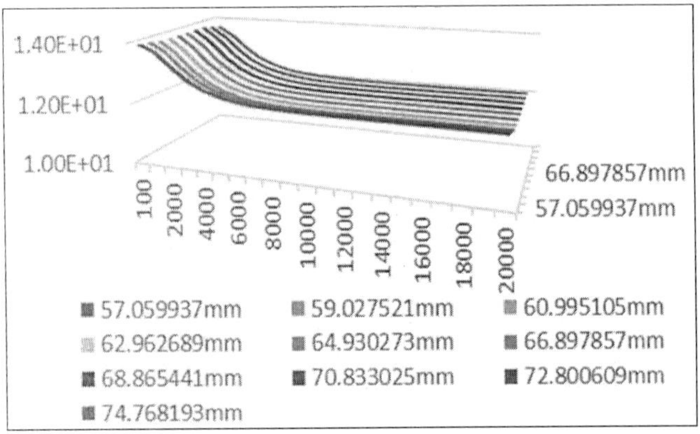

Fig 5.19: 3-D plot of stress, time vs distance at concave side

Convex side of epoxy

From the below figure, the Maximum Electric Stress obtained is 26.5 KV/mm.

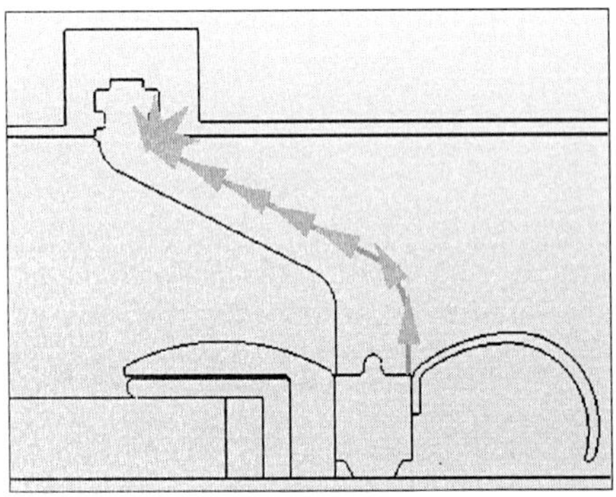

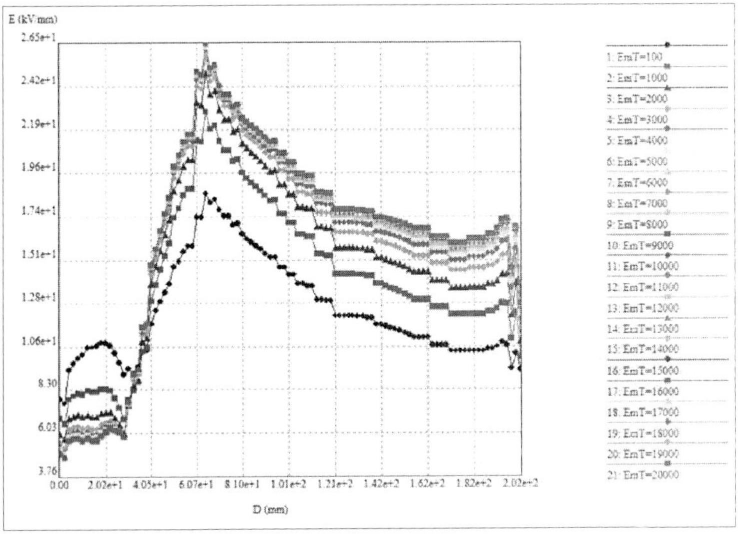

Fig 5.20: E-field result of convex side of cone insulator

Representation of electric field stress vs time of convex side of cone insulator

Table 5.7

Time	Stress
100	1.86E+01
500	2.10E+01
1000	2.24E+01
2000	2.42E+01
3000	2.51E+01
4000	2.55E+01
5000	2.56E+01
6000	2.57E+01
7000	2.58E+01
8000	2.58E+01
9000	2.58E+01
10000	2.65E+01
11000	2.65E+01
12000	2.65E+01
13000	2.65E+01
14000	2.65E+01
15000	2.65E+01

16000	2.65E+01
17000	2.65E+01
18000	2.65E+01
19000	2.65E+01
20000	2.65E+01

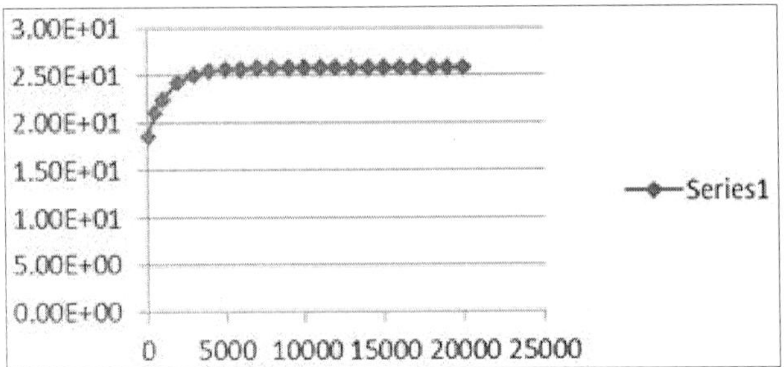

Fig 5.21: Plot between Time vs Stress for convex side of cone insulator

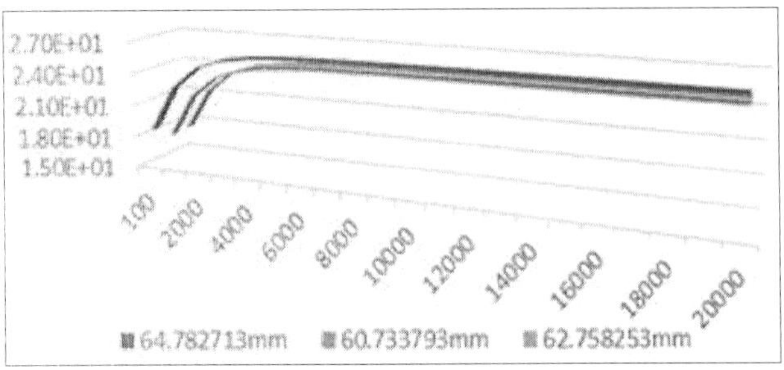

Fig 5.22: 3-D plot of stress, Time vs Distance at convex side

5.3 Comparison of electric field stress with NO insulator, post insulator and cone insulator

5.3.1 Steady state analysis

Below graph shows the comparison of electric field stress with NO insulator, post insulator and cone insulator using static analysis. By observing the below graph, we can say that No insulator and post insulator has maximum stress compare to cone insulator. Therefore, we can conclude that the electric field stress across the cone insulator is optimum and hence cone insulator is preferred over post insulator.

Table 5.8

Insulator condition	Stress (KV/mm)
No Insulator	2.55E+01 KV/mm
Post Insulator	1.67E+01 KV/mm
Cone Insulator	1.40E+01 KV/mm

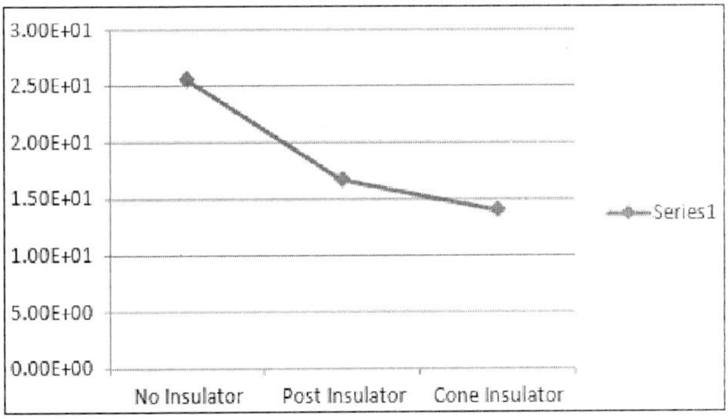

Fig 5.23: Comparison of stress for static analysis

5.3.2 Transient analysis

Below graph shows the comparison of electric field stress with NO insulator, post insulator and cone insulator using transient analysis. By observing the below graph, we can say that No insulator and post insulator has maximum stress compare to cone insulator. Therefore, we can conclude that the electric field stress across the cone insulator is optimum and hence cone insulator is preferred over post insulator.

Table 5.9

Insulator condition	Stress (KV/mm)
No Insulator	3.76E+01 KV/mm
Post Insulator	1.61E+01 KV/mm
Cone insulator	1.30E+01 KV/mm

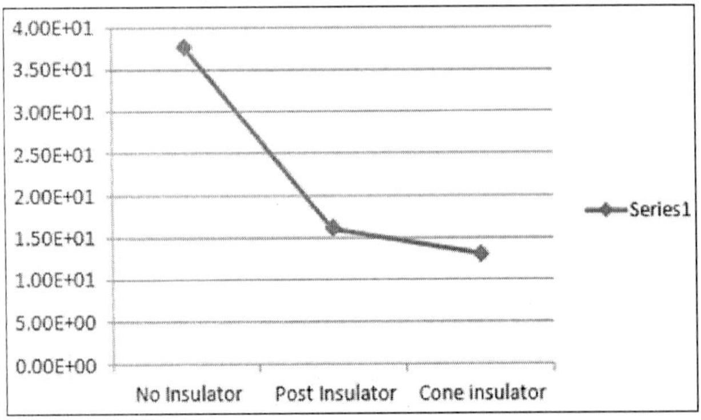

Fig 5.24: Comparison of stress for transient analysis

5.4 Comparison of static and transient analysis

Below graph shows the comparison of Electric field stress in Static and Transient analysis. By observing the below graph, we can say that for the cone insulator in Static (AC analysis) has maximum stress compare to Transient (DC analysis). Therefore, we can conclude that the electric field stress in Transient analysis is optimum and hence DC analysis is preferred over AC analysis for cone insulator.

Table 5.10

Insulator condition	Static stress (KV/mm)	Transient stress (KV/mm)
No Insulator	25.5	37.6
Post Insulator	16.7	16.1
Cone Insulator	14	13

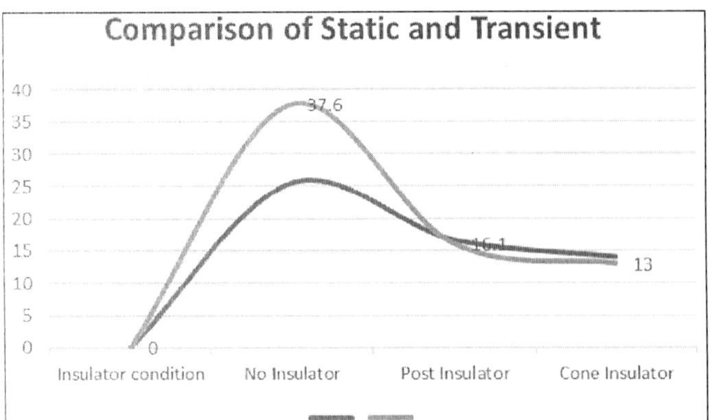

Fig 5.25: Comparison of stress in static and transient

Table 5.11: Maximum electric stress for convex side of cone insulator with different conductivity of epoxy

Conductivity of epoxy (S/m)	Max electric stress (KV/mm)	Distance of epoxy (mm)
6.6*10^-10	25.2	68.9
6.6*10^-12	25.3	65.2
6.6*10^-14	26.5	64.6
6.6*10^-16	19.8	62.5
6.6*10^-18	19.3	62.5
6.6*10^-20	19.3	62.5
6.6*10^-22	19.3	62.5

Table – 5.12 6 Maximum electric stress for concave side of cone insulator with different conductivity of epoxy

Conductivity of epoxy (S/M)	Max electric stress (KV/mm)	Distance of epoxy (mm)
6.6*10^-10	12.2 KV/mm	59 mm
6.6*10^-12	13.8 KV/mm	59 mm
6.6*10^-14	13.9 KV/mm	59.05 mm
6.6*10^-16	13.9 KV/mm	40 mm
6.6*10^-18	13.8 KV/mm	40.1 mm
6.6*10^-20	13.01 KV/mm	42 mm
6.6*10^-22	13.01 KV/mm	42.9 mm

Chapter - 6

Surface Charge Accumulation in HVDC

6.1 Introduction

Surface charge accumulation is correlated with insulator and insulating gas electrical conductivity, and temperature has a significant impact on electrical conductivity. Therefore, it is important to consider how temperature affects the buildup of surface charges while the GIL is in use. Surface tension is assumed to build up under the strain of the electric field because of the discontinuity and nonlinearity in the conductivity, field emission caused by the points on the electrode surface, partial discharge caused by the air gap between the solid dielectric and electrode, and interlayer polarisation considering the difference between the dielectrics. Additionally, under the HVDC electric field, the buildup of surface charge would be a significant problem.

When transmission line is running under HVDC condition i.e., in HVDC GIS when an insulator is subjected to high electric stress, space charge accumulation can induce local increases of the electric field, which must be known and considered for the equipment design when the fields exceed several KV/mm. When the insulator was standing on a grounded plane, the effect of surface charging was dependent on the polarity of the applied impulse voltage. In the case of positive impulse, presence of positive charge increased the flashover voltage, whereas negative charge yielded their reduction and vice versa.

6.2 Factors that affect surface charge accumulation

The surface flashover will almost certainly be caused by the accumulation of surface charges. Effects of surface charge buildup from six variables on the surface charging property of insulators, including insulator-electrode shape, surface roughness of the insulator and conductor, metal particles, temperature, humidity, and gas type.

6.3 Analysis of surface charge accumulation

GIS compact design has steadily gained industry acceptance in recent years. Manufacturers of GIS are paying increasingly greater attention to the insulation performance of cone-type spacers, particularly the insulation qualities of cone-type insulators, as compact designs become more popular. In this project, one such cone insulator is explored. An analysis of the electric field is done using an impulse voltage of 1300 KV. Effect of surface charge for 350 kV GIS spacer with this superimposed voltage is analysed at different positions. Here we are considered 5 positions which are nearer to cone epoxy insulator i.e., concave side of epoxy and convex side of epoxy and applying the total charge of ($1e^{-09}$ C) and run the design for time period of 5.5hrs (Software used: ELECTRO 2D). Effect of surface charge accumulation under different DC electric stress is calculated.

6.4 Methodology

6.4.1 2D model creation in Electro

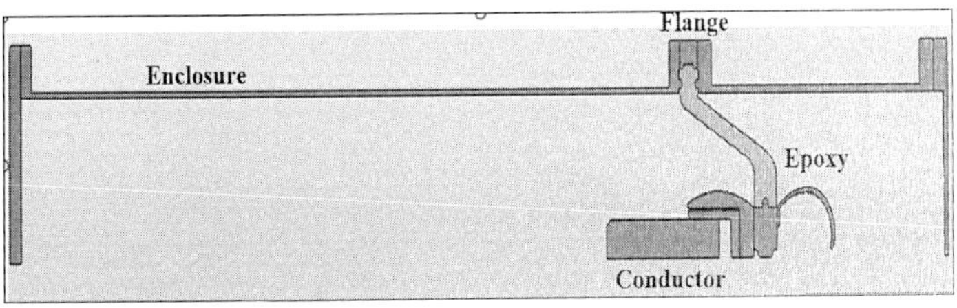

Fig 6.1: Cone epoxy spacer using HVDC analysis

6.4.2 Defining material and boundary condition

Geometry was created as shown in the above figure and uniform boundary conditions was applied. Material is defined for enclosure, conductor and insulator as shown in the figure. The material used for enclosure and conductor is pure aluminium and for insulator is standard moulded epoxy. Uniform voltage was applied for conductor and enclosure, for conductor high potential 1300 KV is applied and for enclosure zero voltage is applied.

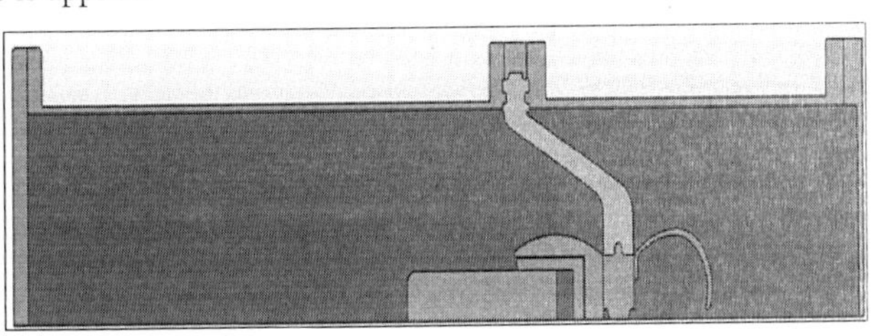

Fig 6.2: Assigning material properties to the cone insulator

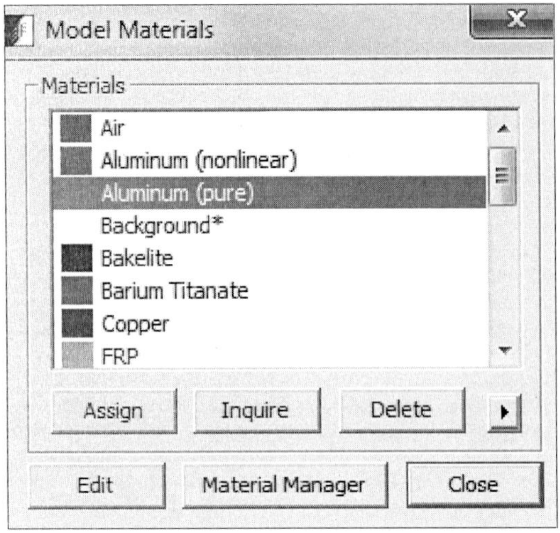
Model Materials
Materials
Air
Aluminum (nonlinear)
Aluminum (pure)
Background*
Bakelite
Barium Titanate
Copper
FRP
Assign
Inquire
Delete
Edit
Material Manager
Close

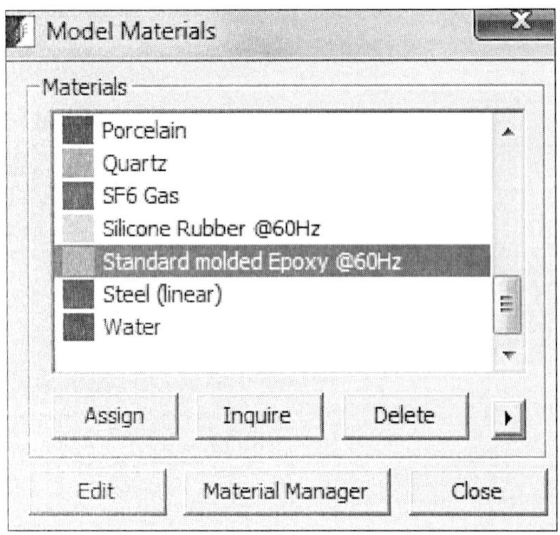
Model Materials
Materials
Porcelain
Quartz
SF6 Gas
Silicone Rubber @60Hz
Standard molded Epoxy @60Hz
Steel (linear)
Water
Assign
Inquire
Delete
Edit
Material Manager
Close

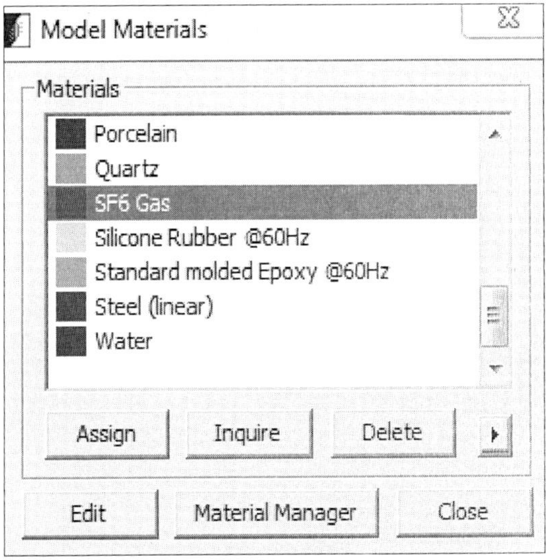

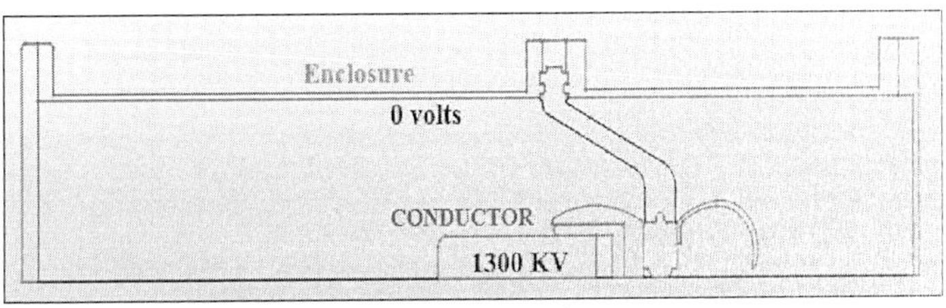

Fig 6.3: Assigning uniform voltage boundary conditions to the dielectric region

6.4.3 Defining element and simulation run

The electric conductivity of aluminium considered for present study is 36900000 S/m and conductivity of Epoxy as 6.6e^{-14} S/m. The conductivity of aluminum components is constant and only the conductivity values for Epoxy was varied to study the effect of various conductivities on the insulator over a period of time. Time period of 5.5hrs i.e., 20,000sec to check the field Analysis Result graph along segment for the time period. After assigning all the required values the Simulation was run and results were studied. Here we are considered 5 positions which are nearer to cone epoxy insulator i.e., concave side of epoxy and convex side of epoxy and applying the total charge of (1e^{-09} C) and run the design for time period of 5.5hrs (Software used: ELECTRO 2D). Under various DC electric stresses, the impact of surface charge accumulation is calculated.

6.5 Results of surface charge accumulation

Position 1: Electric field distribution along the surface of the insulator for the applied surface charge at position1. Electric stress obtained on cone insulator is 28.9 KV/mm.

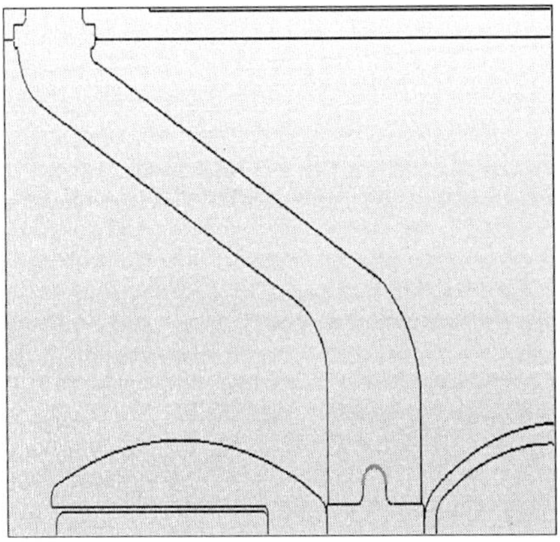

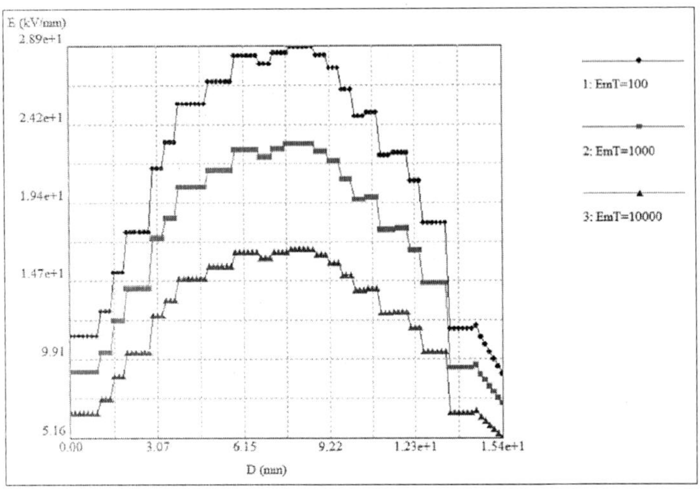

Fig 6.4: E-field stress result on surface charge at position-1

Position 2: Electric field distribution along the surface of the insulator for the applied surface charge at position2. Electric stress obtained at High voltage side and near to convex side of cone insulator as shown in the position 2 is 40.7 KV/mm.

Surface charge at HT side is creating more stress i.e., it has more impact on compare to LV side.

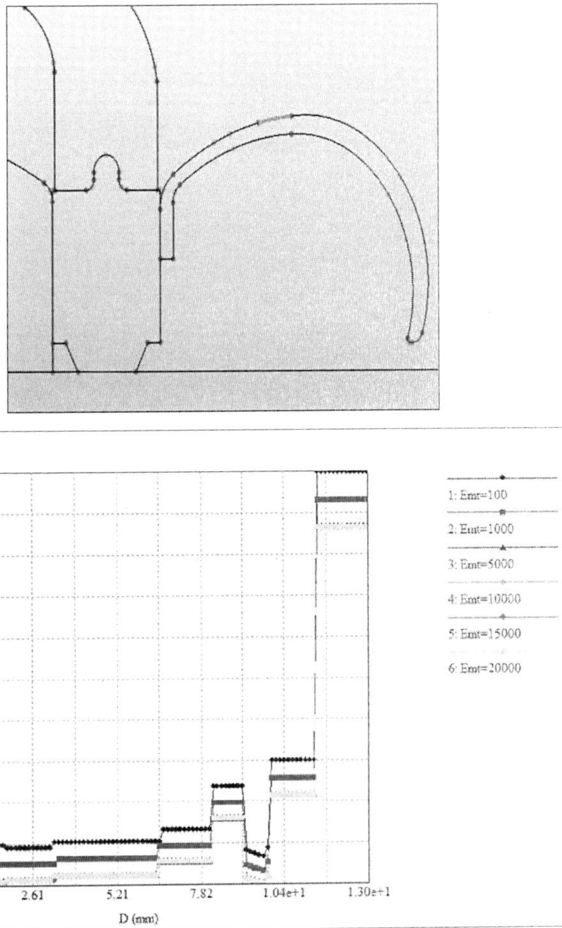

Fig 6.5: E-field stress result on surface charge at position-2

Position 3: Electric field distribution along the surface of the insulator for the applied surface charge at position3. Electric stress obtained at High voltage side and near to concave side of cone insulator as shown in the position 3 is 24.1 KV/mm.

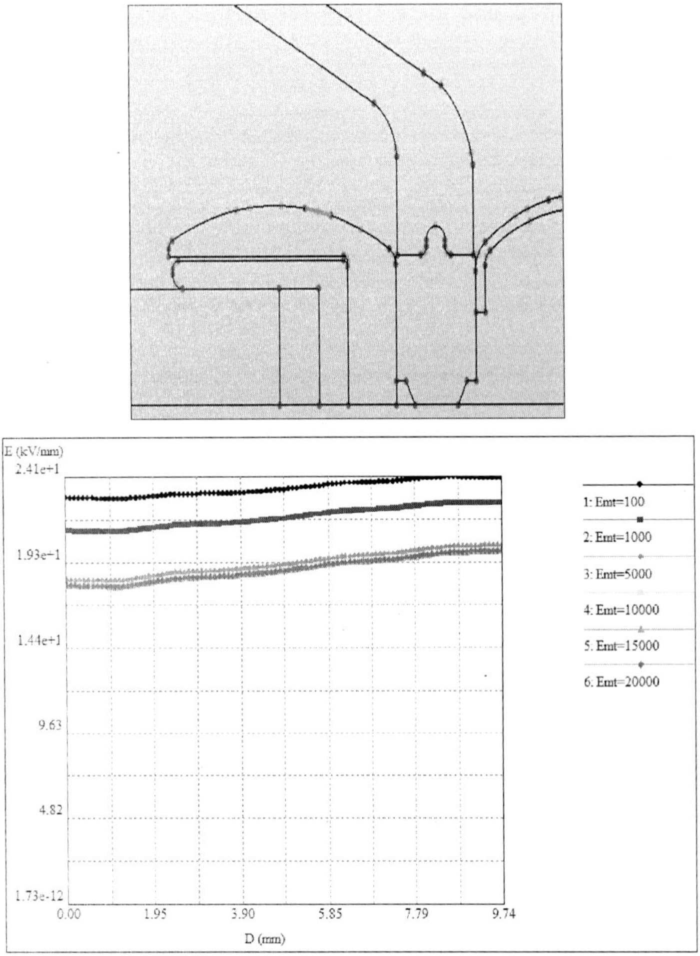

Fig 6.6: E-field stress result on surface charge at position-3

Position 4: Electric field distribution along the surface of the insulator for the applied surface charge at position1. Electric stress obtained at Low voltage side nearer to concave side of cone insulator as shown in the position 4 is 5.45 KV/mm.

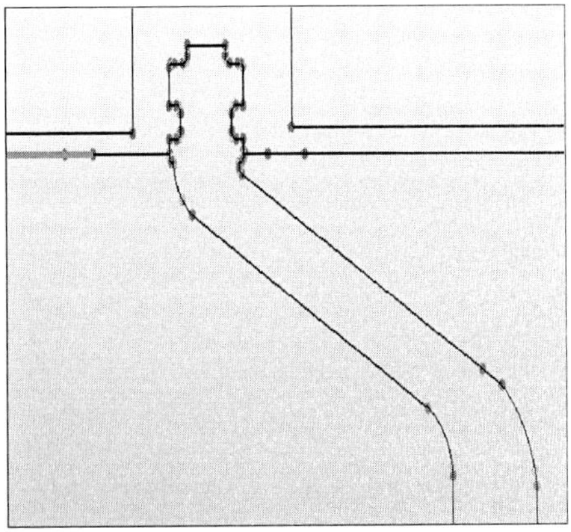

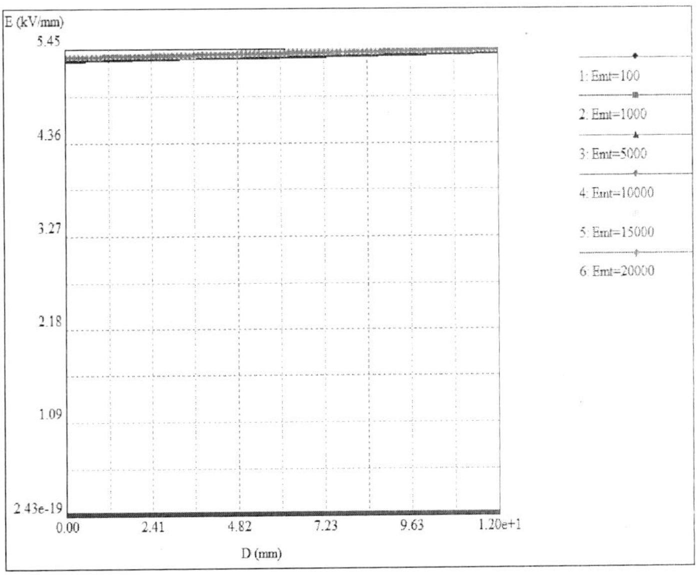

Fig 6.7: E-field stress result on surface charge at position-4

Position 5: Electric field distribution along the surface of the insulator for the applied surface charge at position 5. Electric stress obtained at Low voltage side nearer to convex side of cone insulator as shown in the position 5 is 15.8 KV/mm.

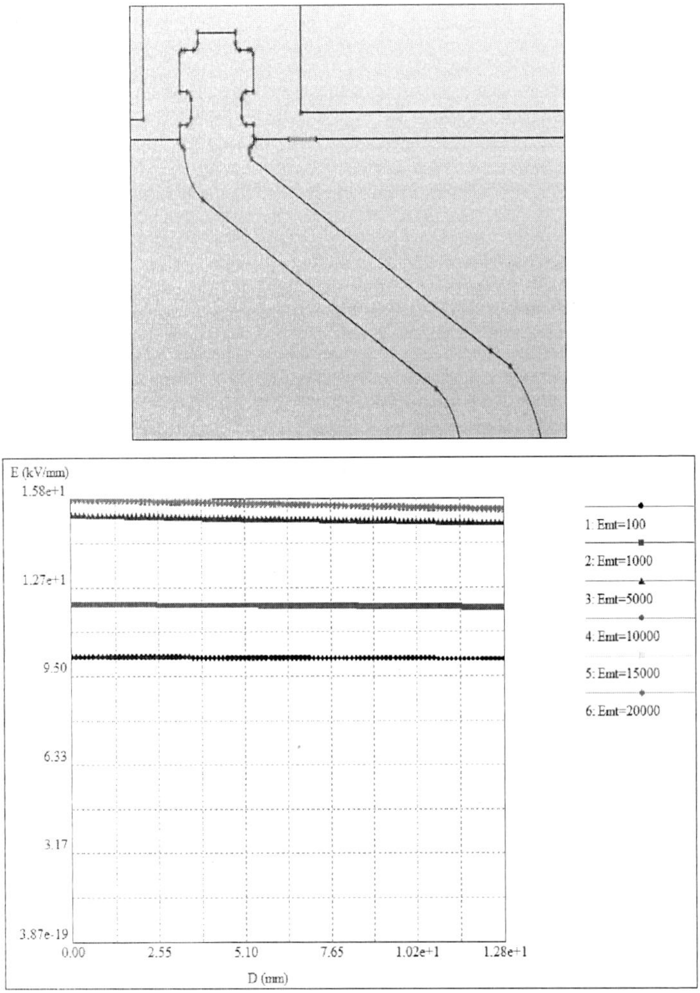

Fig 6.8: E-field stress result on surface charge at position-5

Conclusion: From the above figures at different positions we got different stress. It has noticed that surface charge when accumulated on the HT side nearer to the convex side of the cone insulator , it has more detrimental effect than the all different positions.

Chapter - 7

Conclusions

Following conclusions were made:

1. Comparison between no insulator, post insulator and insulator with certain geometrical profile as discussed in the paper shows that the electrostatic field stress is directly related to geometry of the insulator. For supporting the conductor and for high mechanical strength and to not occur flashovers in HVDC GIS, Insulator is important to employ.

2. From analysis of no insulator, post insulator and cone insulator notice that no insulator has maximum stress i.e., 3.76e+1 KV/mm compare to post and cone insulator. And we can notice that as the profile of the insulator changes the maximum electrostatic stress reduces.

3. Compare graphs of AC and transient analysis we can notice at which instant of time the maximum stress in AC analysis coordinate with the stress in transient analysis.

   i) At=100 sec on concave side of epoxy under ransient analysis is matching with the maximum stress in AC analysis i.e., 14 KV/mm.

   ii) At t= 1000 sec on convex side of epoxy under ransient analysis is matching with the maximum stress in AC analysis i.e., 20.7 KV/mm.

4. Transient analysis is carried out for various conductivities it was found that there is no variation of electrostatic field stress for the conductivities of range 6.6e-10 S/m to 6.6e-12 S/m and 6.6e-16 S/m to 6.6e-22 S/m but with the conductivities of range 6.6e-12 S/m to 6.6e-14 S/m noticed that there are multiple variations of stress. By noticing the results of electrostatic field stress, we can observe that convex side has maximum stress compare convex side for cone epoxy spacer.

5. Space charge accumulation can cause local increases in the electric field when an insulator is subjected to high electric DC stress when a transmission line is operating under HVDC conditions, which must be known and considered for the equipment design when the fields exceed several KV/mm.

6. To understand the effect of space charge on the insulator, the space charge is positioned at four different places and the effect was seen, it is seen that

   a) If the space charge (1e-09) is established at High voltage side and near to convex side then the stress obtained is 40.7 KV/mm.

   b) If the space charge (1e-09) is established at High voltage side and near to concave side then the stress obtained is 24.1 KV/mm.

   c) If the space charge (1e-09) is established at Low voltage side and near to concave side then the stress obtained is 5.45 KV/mm.

   d) If the space charge (1e-09) is established at Low voltage side and near to convex side then the stress obtained is 15.8 KV/mm.

7. Hence wecan conclude that surface charge when accumulated on the HT side nearer to the convex side of the cone insulator, it has more detrimental effect than the all different positions.

8. Comparison of electric field stress with NO insulator, post insulator and cone insulator we can notice that No insulator and post insulator has maximum stress compare to cone insulator. Therefore, we can conclude that the electric field stress across the cone insulator is optimum and hence cone insulator is preferred over post insulator.

9. Comparison of Electric field stress in Static and Transient analysis we can notice that for the cone insulator the Static (AC analysis) has maximum stress compare to Transient (DC analysis). Therefore, we can conclude that the electric field stress in Transient analysis is optimum and hence DC analysis is preferred over AC analysis for cone insulator.

References

1. Mitigation of field stress with metal inserts for cone type spacer in a gas insulated busduct under delamination. https://www.sciencedirect.com/science/article/pii/S2215098616312137.

2. Optimal Design of Spacer for High Voltage Gas Insulated Switchgear HyunWoo Joo, ChaeYoon Bae, JongUng Choi, YoungGeun Kim LS Industrial Systems/Electromechanical R&D Center 68, Wolmyeong-ro #201, Heung deok-gu, Cheongju-si, Chungcheongbuk-do, Korea, 361-817.

3. Haoran Wang, Zihao Guo, Hua Feng, Lilan Liu, Huidong Tian Chuang Wang and Zongren Peng, "Simulation and Experimental Study on Mechanical Performance of UHV GIS Spacer in Hydrostatic Test," IEEE Transactions on Dielectrics and Electrical Insulation Vol. 24, No. 5., pp3122-3131.

4. Evgeni Volpov, "Dielectric Strength Coordination and Generalized Spacer Design Rules for HVACrDC SF Gas Insulated Systems", IEEE Transactions on Dielectrics and Electrical Insulation, Vol. 11, No. 6; December 2004, pp949-963.

5. Hisashi Goshima, Shigemitsu Okabe, Toshiaki Ueda, Hiroshi Morii, Nobuhiko Yamachim, Koji Takahata, and Masayuki Hikita, "Fundamental Insulation Characteristics of High-Pressure CO_2 Gas for Gas-Insulated Power Equipment", IEEE Transactions on Dielectrics and Electrical Insulation Vol. 15, No. 4, pp1023-1030.

6. Syarif Hidayat and Fransiskus Damanik, "Electric Field Optimization on 150 kV GIS Spacer by Modification of Spacer Shape and Conductor Configuration" The 3rd IEEE Conference on Power Engineering and Renewable Energy, ICPERE 2016, pp225-230.

7. https://library.e.abb.com/public/7cba8707e4e849eab2831c2a211c9695/Compact%20Gas-insulated%20Systems%20for%20High%20Voltage%20Direct%20Current%20Transmission_Basic%20Design.pdf.

8. Dielectric Strength Coordination and Generalized Spacer Design Rules for HVACrDC SF Gas Insulated Systems 6 Evgeni Volpov Planning, Development & Technology Division, Israel Electric Corporation Ltd. Haifa 31000, Israel.

9. https://www.researchgate.net/publication/361564757_Surface_Flashover_Characteristics_of_Epoxy_Resin_Composites_in_SF6CF4_Gas_Mixture_with_DC_Voltage.
10. K. Nakanishi, A. Yoshioka, Y. Arahata and Y. Shibuya, ''Surface Charging on Epoxy Spacer at DC Stress in Compressed SF gas'', 6 IEEE Trans. Power App. Syst., Vol. 102, pp. 3919 3927, 1983.
11. V. Varivodov and E. Volpov, ''An Approach to the Spacer Design of HVDC SF Gas Insulated Equipment'', 7th ISH, Dres- 6 den, Germany, paper 31.09, 1991.
12. E. Volpov, ''HVDC Gas Insulated Apparatus: Electric Field Specificity and Insulation Design Concept'', IEEE Electr. Insul. Mag., Vol. 18, No 2, pp. 7 14, 2002.
13. F. Messerer; W. Boeck Field optimization of an HVDC-GIS-spacer.
14. E. Volpov, "Dielectric strength coordination and generalized spacer design rules for HVAC/DC SF 6, gas insulated systems", IEEE Trans. Dielectr. Electr. Insul, Vol. 11, pp. 949-963, 2004.
15. B. Mobasher, D. Kingsbury, J. Montesinos and R.S. Gorur, "Mechanical aspects of crimped glass reinforced plastic (GRP) rods", IEEE Trans. Power Delivery, Vol. 18, pp. 852-858, 2003.
16. L.S. Kumosa, M.S. Kumosa and D. L. Armentrout, "Resistance to brittle fracture of glass reinforced polymer composites used in composite (nonceramic) insulators", IEEE Trans. Power Delivery, Vol. 20, pp. 2657-2666, 2003.
17. Y. Chen, B. Cui, N. Wang, Y. Wu, D. Cao and P. Cheng, "Structural Design of Central Insert in Basin-type Insulator Used for 1100 kV GIS", High Voltage Eng., Vol. 42, pp. 564-570, 2016 (in Chinese).
18. Y. Chen, B. Cui, N. Wang, Y. Wu, D. Cao and P. Cheng, "Study of Improving Mechanical Strength for 1100 kV Spacer Insulators", High Voltage Apparatus, Vol. 51, pp. 52-56, 2015
19. "Insulation characteristics of epoxy insulator with internal delamination shaped micro-defects", IEEE Trans. Dielectr. Electr. Insul, Vol. 20, pp. 1851-1858, 2013.
20. G. Ueta, J. Wada, S. Okabe, M. Miyashita, C. Nishida and M. Kamei, "Insulation characteristics of epoxy insulator with internal crack-shaped micro-defects-fundamental study on breakdown mechanism", IEEE Trans. Dielectr. Electr. Insul, Vol. 20, pp. 1444-1451, 2013.

21. H. Wang, Z. Peng, Z. Guo, Q. Wang, J. Liao and S. Zhang, "Simulation study of electric influence caused by defects on UHV AC GIS spacer", IEEE Conf. Properties and Applications of Dielectric Materials, pp. 676-679, 2015.